Heinz Gaderer, Johannes Zuber,
Josef Fischer, Charlotte Harrer-Kirnbauer, H. E. DeGrear

Working on Metals

English for Mechanical Engineering Staff

**Ein Lern- und Arbeitsbuch
für den handlungsorientierten Englischunterricht**

This book belongs to

first name(s)

surname

form

school

home address

address at work

Internet

Sie finden uns im Internet unter: http://www.stam.de

Stam Verlag
Fuggerstraße 7 · 51149 Köln

ISBN 3-8237-**4823**-8

Dieses Schulbuch wurde entwickelt in Zusammenarbeit mit dem Österreichischen Gewerbeverlag Ges.m.b.H., Wien.

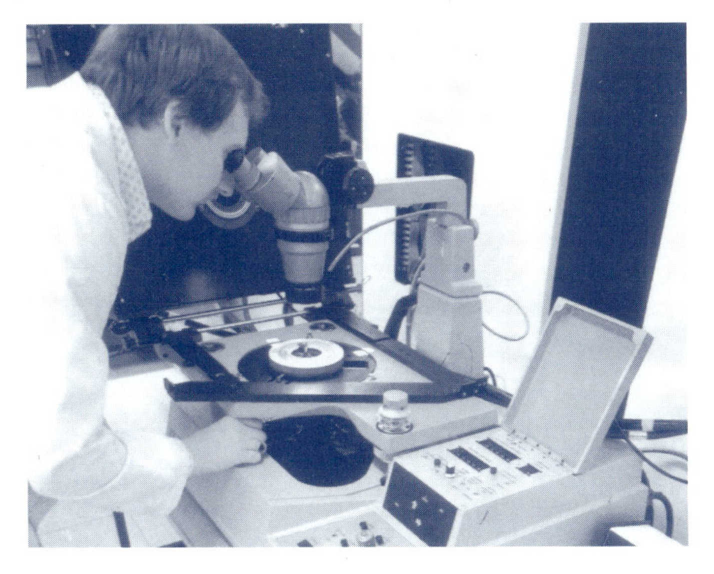

Les Smith
The motor mart for every part
OPEN
SUNDAYS
OPEN ON SUNDAYS
No.1
£2.99
£89.99
OPEN SUNDAYS
The motor mar
WD-40
WE STOCK
BLACK+DECKER
ROOF RACKS
DOG GUARDS
CAR ALARMS
IN CAR HIFI
SEAT BELTS
WAX N WASH
CHILD SEATS
SEAT COVERS
OPEN - 6PM

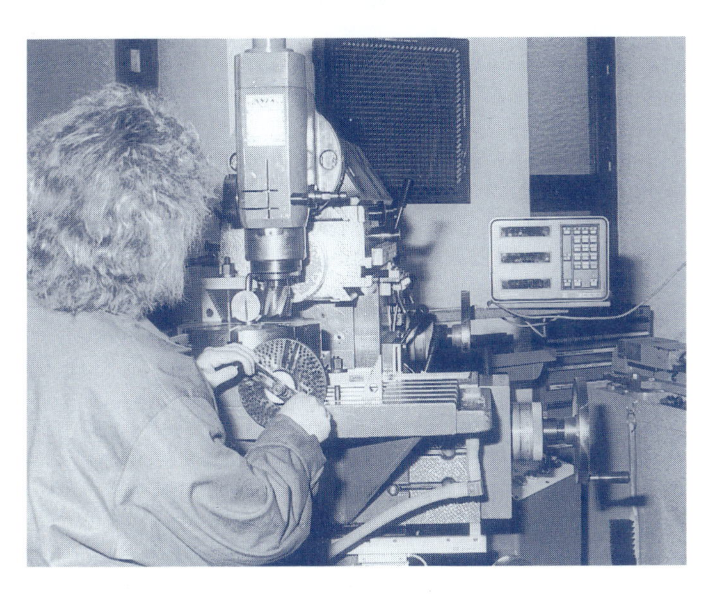

Zu **Working on Metals** gibt es eine **Tonkassette** mit Hörspielen, Dialogen, Nachrichten und Liedern zum Üben des Hörverstehens und als Vorlage für das Sprechen.
Man kann sie auch zu Hause hören:
Working on Metals ISBN 3-8237-4823-8
Speakers: Buky Lardner and Howard Nightingall from England,
Amy Lindner, Kathryn Lucius and Dennis Kozeluh
from the USA.
Recorded at Soundborn Studios Vienna

Contents

Contents

Contents

Dear Student,

Bevor Sie mit der eigentlichen Arbeit beginnen, möchten wir Ihnen Ihr Lehr- und Arbeitsbuch vorstellen. **Working on Metals** soll Sie auf Begegnungen mit der englischen Sprache vorbereiten: an Ihrem Arbeitsplatz, an der Schule, in Ihrer Freizeit und auf Reisen.

Um sie möglichst wirklichkeitsnah an die englische Sprache heranzuführen, haben wir uns jeweils an authentischen Texten und Bildern aus englischen Prospekten, Broschüren, Büchern, Zeitungen und Fachzeitschriften orientiert. In zahlreichen Gesprächen mit Ihren Berufskollegen in England haben wir uns eingehend über Maschinenbau- und metallverarbeitende Berufe informiert. Die Fachgespräche und Interviews dienten als Vorlage für die Tonkassetten zu **Working on Metals.** Authentische Texte sind also Hör- und Lesetexte, wie sie im wirklichen Leben vorkommen.

Diese Texte haben wir für den Englischunterricht an berufsbildenden Schulen bearbeitet und dabei versucht, auf Ihre Erwartungen und Interessen einzugehen und Ihre Anregungen aufzugreifen. Das hat uns sehr geholfen. Und damit wären wir auch schon beim Lernen: So wie wir mit Kollegen in England und an unseren Schulen zusammengearbeitet haben, können auch Sie mit Ihrem Wissen, Ihren Vorkenntnissen und mit Ihren ersten beruflichen Eindrücken zum Gelingen des Englischunterrichts beitragen.

Wollen Sie sich gleich einmal überlegen, mit welchen Themen und Texten Sie beruflich zu tun haben (werden)?
Setzen Sie sich in kleinen Gruppen zusammen und reden Sie miteinander darüber. Wenn Sie Ihre Ideen auf einen Bogen Papier schreiben, können Sie ihn an die Wand hängen und mit anderen Gruppen vergleichen und ergänzen.

Dann können Sie Ihre Poster mit dem Inhaltsverzeichnis vergleichen. Dort finden Sie Themen und Texte, die wir für Sie ausgewählt haben, gleich in Englisch. Wie ist der Vergleich ausgefallen?

Der Umgang mit englischen Themen und Texten verlangt bestimmte Fertigkeiten, die uns in der Muttersprache selbstverständlich sind, auf die es beim Erlernen einer Fremdsprache aber besonders ankommt. Welche Fertigkeiten finden Sie beim Englischlernen für Ihren Beruf am wichtigsten?
Bitte reihen Sie sie von 1 (am wichtigsten) bis 5 (am wenigsten wichtig).

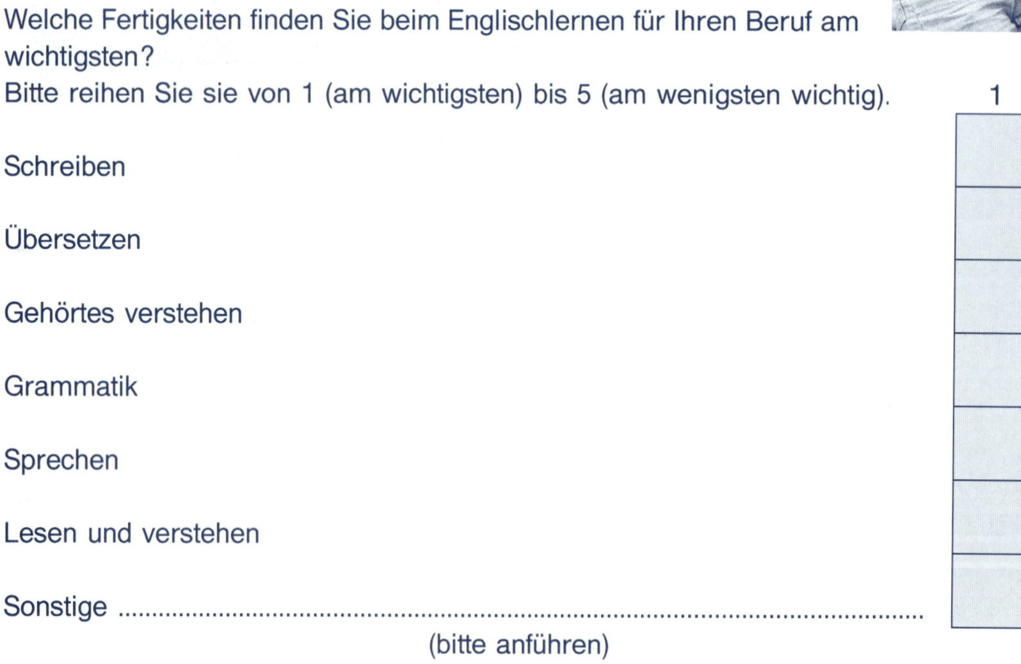

	1	2	3	4	5
Schreiben					
Übersetzen					
Gehörtes verstehen					
Grammatik					
Sprechen					
Lesen und verstehen					
Sonstige .... (bitte anführen)					

Vergleichen Sie Ihr Ergebnis mit den Kollegen in Ihrer Gruppe und diskutieren Sie das Gesamtergebnis in Ihrer Klasse.

Vielleicht möchten Sie an dieser Stelle auch besprechen, was Ihnen an Ihrem Englischunterricht bisher besonders gefallen hat? Und was weniger gut angekommen ist?

Was erwarten Sie sich deshalb vom Englischunterricht an Ihrer Schule?
Und von einem Lehrbuch?

Die Arbeit mit **Working on Metals** soll Ihnen helfen, Ihre Interessen und Erwartungen zu verwirklichen. Daher möchten wir Ihnen vorschlagen, wie Sie mit unserem Buch arbeiten können. Jede Unit (Lerneinheit) enthält ausreichend Texte und Arbeitsvorschläge für jeweils eine Unterrichtseinheit. Die einzelnen Übungen sind zwar aufeinander abgestimmt und aufbauend, doch brauchen Sie nicht unbedingt immer alle durchzuführen. Viel wichtiger erscheint uns, daß Sie sich intensiv mit den Texten beschäftigen und ausreichend zum Sprechen kommen. Und dabei würde uns vor allem interessieren, **was** Sie zu den Themen sagen möchten. Vielleicht interessiert es Sie manchmal auch, die Tonaufnahmen und Lesetexte danach zu untersuchen, wie sich die Engländer sprachlich **richtig** ausdrücken.
Mit der Zeit werden Sie die von uns vorgeschlagenen Übungstypen so gut kennenlernen, daß Sie jene auswählen können, die Ihnen am besten gefallen und mit denen Sie am besten lernen können. Je nach Lehrberuf oder Fachrichtung wird es erforderlich sein, die für Sie am besten geeigneten Themen auszuwählen oder Texte zum Unterricht beizusteuern, mit denen Sie in Ihrem Beruf zu tun haben (werden). Selbstverständlich werden Sie ihre Entscheidungen nicht allein treffen. Ihr Lehrer oder Ihre Lehrerin wird Ihnen geeignete Vorgangsweisen empfehlen; er/sie wird Sie gern in allen Fragen, die beim Lernen auftreten, beraten und unterstützen.

Reden Sie auch mit Ihren Kollegen und mit Ihrem Lehrer oder mit Ihrer Lehrerin darüber, wie Ihnen die Arbeit mit **Working on Metals** gefällt.

Das Lehrbuch soll Sie anregen, selbständig zu arbeiten und miteinander zu lernen, mit Ihren Kollegen und Lehrern ins Gespräch zu kommen und einander in der englischen Sprache zu begegnen. Die Arbeit mit **Working on Metals** wird Ihnen hoffentlich auch Spaß machen.

Wir würden uns freuen, von Ihnen zu hören. Schreiben Sie uns einfach, vielleicht sogar in Englisch?
Have fun with **Working on Metals!**

Heinz Gaderer, Josef Fischer, Charlotte Harrer-Kirnbauer, Johannes Zuber

1 Meet our speakers

Look at the photo on the left and listen to the cassette.

Who's who in the photo?
Discuss your result with your neighbour.
What do the other pairs think?

Listen again: this time look at the passport and at the business card.
Tick (✓) the information you hear.
Compare your result with a partner.

Melanie Dalton
VHS Darmstadt

64287 Darmstadt, Mozartweg 10
☎ 06151/71 19 63 VHS: 601 18 - 62 05

→ WARNING: ALTERATION, ADDITION OR MUTILATION OF ENTRIES IS PROHIBITED. ANY UNOFFICIAL CHANGE WILL RENDER THIS PASSPORT INVALID.

NAME—NOM
THOMAS ANTHONY D'AGOSTINO

SEX—SEXE: M
BIRTHPLACE—LIEU DE NAISSANCE: CONNECTICUT, U.S.A.

BIRTH DATE—DATE DE NAISSANCE: AUG. 2, 1950
ISSUE DATE—DATE DE DELIVRANCE: JUNE 14, 1985

NATIONALITY—NATIONALITÉ: UNITED STATES OF AMERICA
EXPIRES ON—EXPIRE LE: JUNE 13, 1995

SIGNATURE OF BEARER—SIGNATURE DU TITULAIRE
NOT VALID UNTIL SIGNED

PHOTOGRAPH ATTACHED
FOREIGN SERVICE
UNITED STATES OF AMERICA

Listen once more:
This time concentrate on information about Marijana Dworski.

Tick the right boxes (✓).

She's
- ○ American ○ from Connecticut.
- ○ Welsh ○ from Wales.
- ○ British ○ from Bournemouth.
- ○ English ○ from England.

Connect the boxes (○—○).

Her first name ○ — is — ○ Polish.
Her surname ○ ○ English.
○ Yugoslav.

Compare your results with a partner.

What can you say about the other speakers?
What nationality are they?
Where are they from?

Melanie ....

Tom ....

Howard ....

Now you have met the speakers on our cassette. You will hear their voices quite often. And you will get to know them better as we go along.

Did you like the exercise?
Did you have any problems?

2 Meet your book

Find answers to the following questions.
Work with a partner.
Make short notes only.

What's the title?

Who are the authors?

Who is the publisher?

When was it published?

How many units are there?

How many pages?

Where is the cassette script?

Anything else you can say?

Hello! I'm your English book
My name is ....

Now fill in **your** personal data on page 1.

3 Meet my partner

Listen to the cassette and fill in the missing words.

.... Marijana Dworski.

Dworski? .... D·W·O·R·S·K·I

Oh, that's nice.

But .... Marijana.

.... years old.
.... English teacher in ....
.... Wales
but .... English.
And .... single.

Would you like to meet one of your classmates?
Talk to him/her and fill in a personal data sheet.
Maybe he or she will give you a photo, too.

name ....

surname ....

age .... marital status ....

occupation .... nationality ....

photo

Now introduce your partner to the others in your group.

4 Who are you?

Write your name on a card.
Introduce yourself like this:

Hello! I'm Andrea.
My name is Andrea Frank.
My friends call me Andie.

You are talking to Andrea "Andie" Frank

Hello, Andie!

Nice to meet you ....

But now I'm going home.
Goodbye.
See you next week.

Goodbye Andie!

Modern metal-working

Jobs in metal-working industries fall into two general categories: Skilled workers (craftsmen, craftswomen) and technical personnel (technicians, engineers).

Skilled workers are found in all areas of metal-working. A few of the speciality fields are welding, sheet metal work, turning, assembly or installation of mechanical machinery, repair and service of automobiles. Many of today's skilled workers receive their training as an **apprentice.** The period of instruction usually requires three or more years of training with experienced craftworkers. In addition to this on the job training an apprentice studies maths, science, English, print reading, metallurgy, safety and production techniques at a vocational school.

Modern technology has brought about a demand for persons capable of doing complex work of a highly technical nature. These men and women are called **technicians.** The technician assists the engineers by constructing and testing experimental devices and equipment. He or she aids in compiling statistical information, making cost estimates and preparing technical reports. Many colleges offer a three year programme. The course of study stresses maths, science, manufacturing and production techniques.

© Wayland

1 Modern metal-working

Are you an **apprentice?**
Or are you going to be a **technician?**

Read the article and mark the passage that concerns you.
Underline the relevant activities in the text.

Look at the list of jobs.
Mark each one with Ⓢ for skilled worker
or Ⓣ for technical staff.

◯ mechanical fitter
◯ draughtsman, draughtswoman
◯ turner
◯ motor mechanic
◯ production technician
◯ quality controller
◯ tool maker
◯ machine fitter
◯ lab(oratory) technician
◯ field tester
◯ welder
◯ sheet metal worker

What does it take to become a **skilled worker** in Great Britain?

And what does it take to become a **technician?**

2 Metalworking jobs

Look at the photos and read the texts on the opposite page.
What do each of the young people do?
What are their jobs?

Susan Sanborn prepares design drawings for new production machinery. She went to a technical college, and is now training to use the computer-aided design system at C. Dugard.

She is a ....

Martin Marek is an apprentice at Haas. He helps build machinery for the food industry. He is learning to read technical drawings and to assemble some of the machines.

He is a ....

Tina Braun just started an apprenticeship at General Motors. She will work on all kinds of engines for automobiles and for production machinery. In a few years she will be

....

Peter Preston studied mechanical engineering at a technical college. He now works for C. Dugard Ltd in Hove, on the south coast of England. He inspects and tests machine tools for quality.

He is a ....

3 What are their jobs?

Look at the photos and listen to the interview on cassette. Who is the interview about?

Where does the person work?

When did he or she start?
How long will the training take?
What is he or she training to be?

Compare your results with a partner.

4 Your career

Interview a partner about his or her career. First complete the questions, then take notes.

What .... training to be?

....

What school .... go to?

....

For how long? ....

When .... start? ....

Where .... work? ....

What .... learn? ....

....

5 Subjects at school

Go through the list of subjects, then listen to the interview again. Tick the subjects that are mentioned.

- ○ maths
- ○ physics
- ○ science
- ○ German
- ○ English
- ○ history
- ○ geography
- ○ technology
- ○ metallurgy
- ○ technical drawing
- ○ political education
- ○ physical education
- ○ print reading
- ○ workshop
- ○ commercial correspondence
- ○ production techniques
- ○ safety measures and first aid

What do you learn at your school or college? Interview a partner, then write a school report for him or her.

A sweet challenge

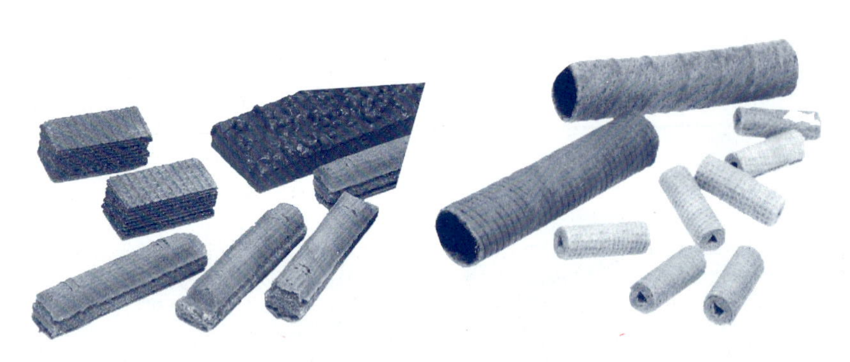

© Haas

cream-filled wafers — wafer sticks and fan wafers — ice cream cones and cups

1 Sweets after all?

Listen to the interview on cassette and answer the following questions.
Tick the correct answer.

Which company does Martin work for?
- ○ General Motors
- ○ Haas
- ○ KTM

What do they produce?
- ○ Wafers
- ○ Sweets
- ○ Machines

What kind of machines have they got there?
- ○ Hand-operated
- ○ Semi-automatic
- ○ Computer-controlled

2 Departments and their fuctions

Look at the photos on the opposite page and read the texts quickly.
It is not important that you understand the texts word for word.
Which text belongs to which photo?

Read the texts again.
What are the fuctions of the three departments?

In design	○	○ they produce wafer machines and spare parts.
In manufacturing	○	○ they improve existing products and develop new products.
In research and development	○	○ they prepare and finish machine configuration plans using CAD-systems.

3 Where do you work?

Which company do you work for?
Which company would you like to work for? ....

What do they produce? ....

What are the products used for? ....

In which department do you/would you like to work? ....

Interview 3 or 4 of your colleagues and take notes.
Then write a short report about the company and department where you (would like to) work.

Our design department uses up-to-date technical aids. Design drawings and machine configuration plans are prepared and finished using an advanced CAD system. We apply the very latest in technology to develop new machines and continuously modernize our existing product line.

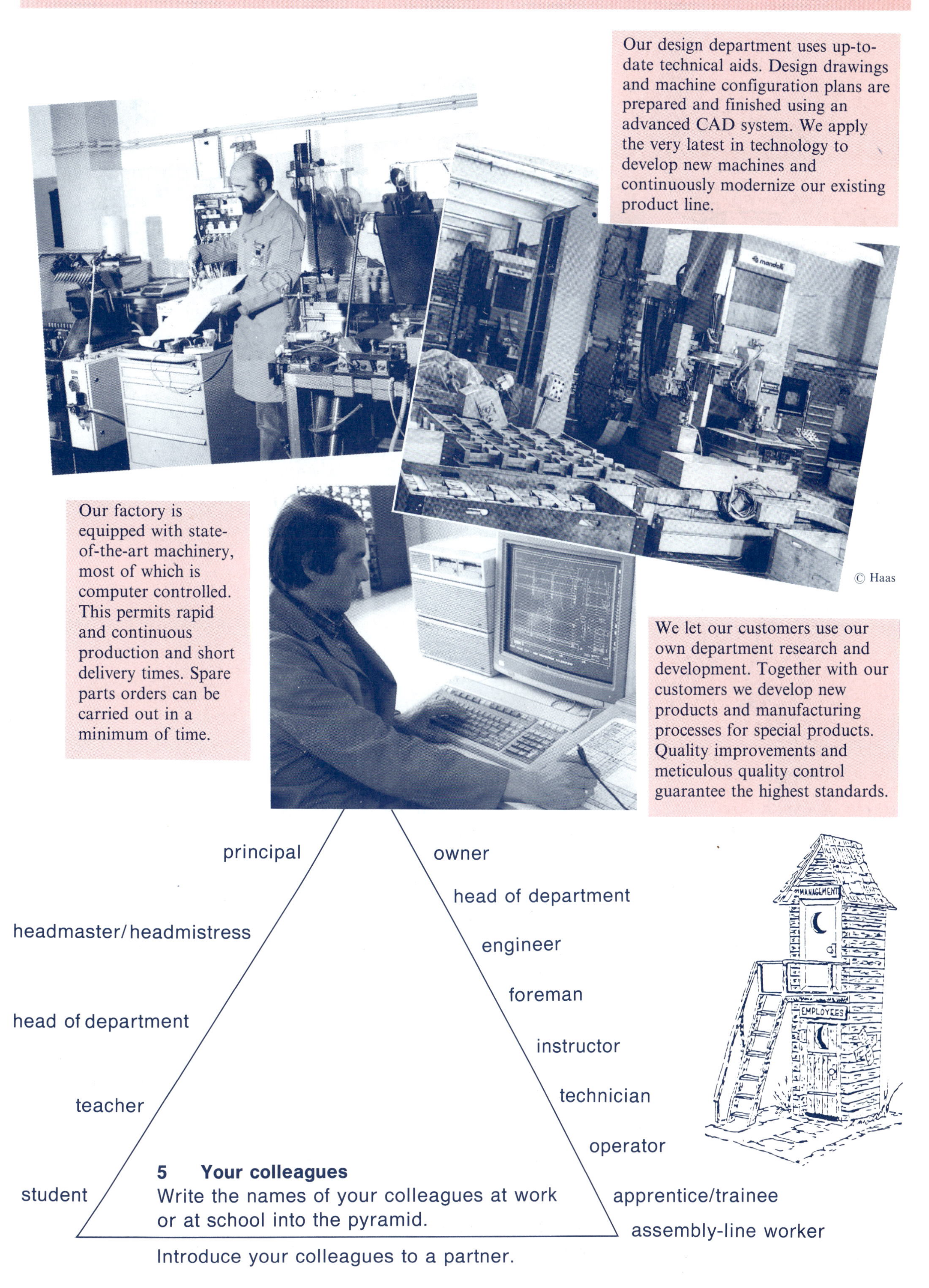

© Haas

Our factory is equipped with state-of-the-art machinery, most of which is computer controlled. This permits rapid and continuous production and short delivery times. Spare parts orders can be carried out in a minimum of time.

We let our customers use our own department research and development. Together with our customers we develop new products and manufacturing processes for special products. Quality improvements and meticulous quality control guarantee the highest standards.

principal
headmaster/headmistress
head of department
teacher
student

owner
head of department
engineer
foreman
instructor
technician
operator
apprentice/trainee
assembly-line worker

5 Your colleagues

Write the names of your colleagues at work or at school into the pyramid.

Introduce your colleagues to a partner.

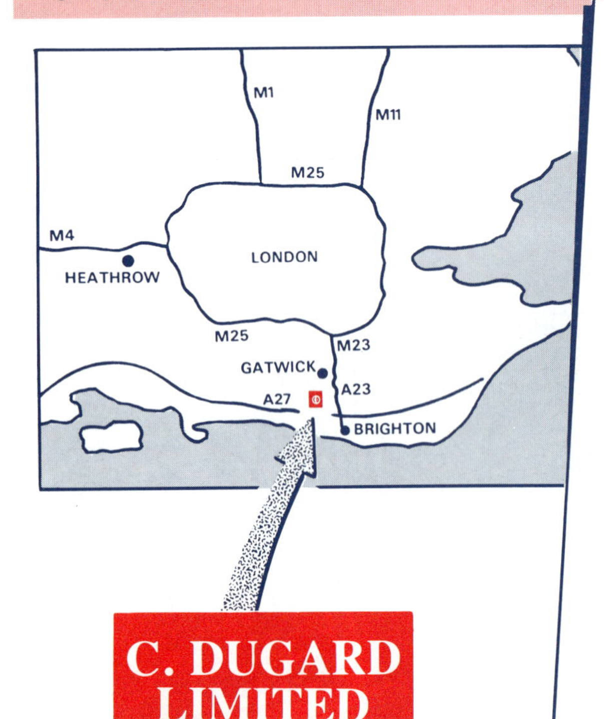

132 Preston Road
Hove, Sussex

16 September 19..

Hello Martin!

You asked me to tell you about my new job. Well, the company is in Hove on Old Shoreham Road. Hove is near Brighton. Look at the map I sent you. The Brighton and Hove Football Club is just across the road. Can you find the company logo? It's CD, for C. Dugard Ltd.

It is a medium size company which produces machine tool centres. On the shop floor there are about fifty technicians, craftsmen and -women all together. We've got mechanics, machine fitters, welders, spraypainters, tool and die makers, computer programmers and quality controllers. They come from all over the world. We sometimes go for a drink together and I have found many new friends.

I would like to know where you work. How do you like your job? Write soon.

Regards, Peter

P.S. If you want to know about the plant layout listen to the cassette I sent you.

C. DUGARD LIMITED
MACHINE TOOLS
75 OLD SHOREHAM ROAD, HOVE, SUSSEX BN3 7BE
Telephone: Brighton (0273) 728581 - 732286
Telex: 877423 Fax: (0273) 203835

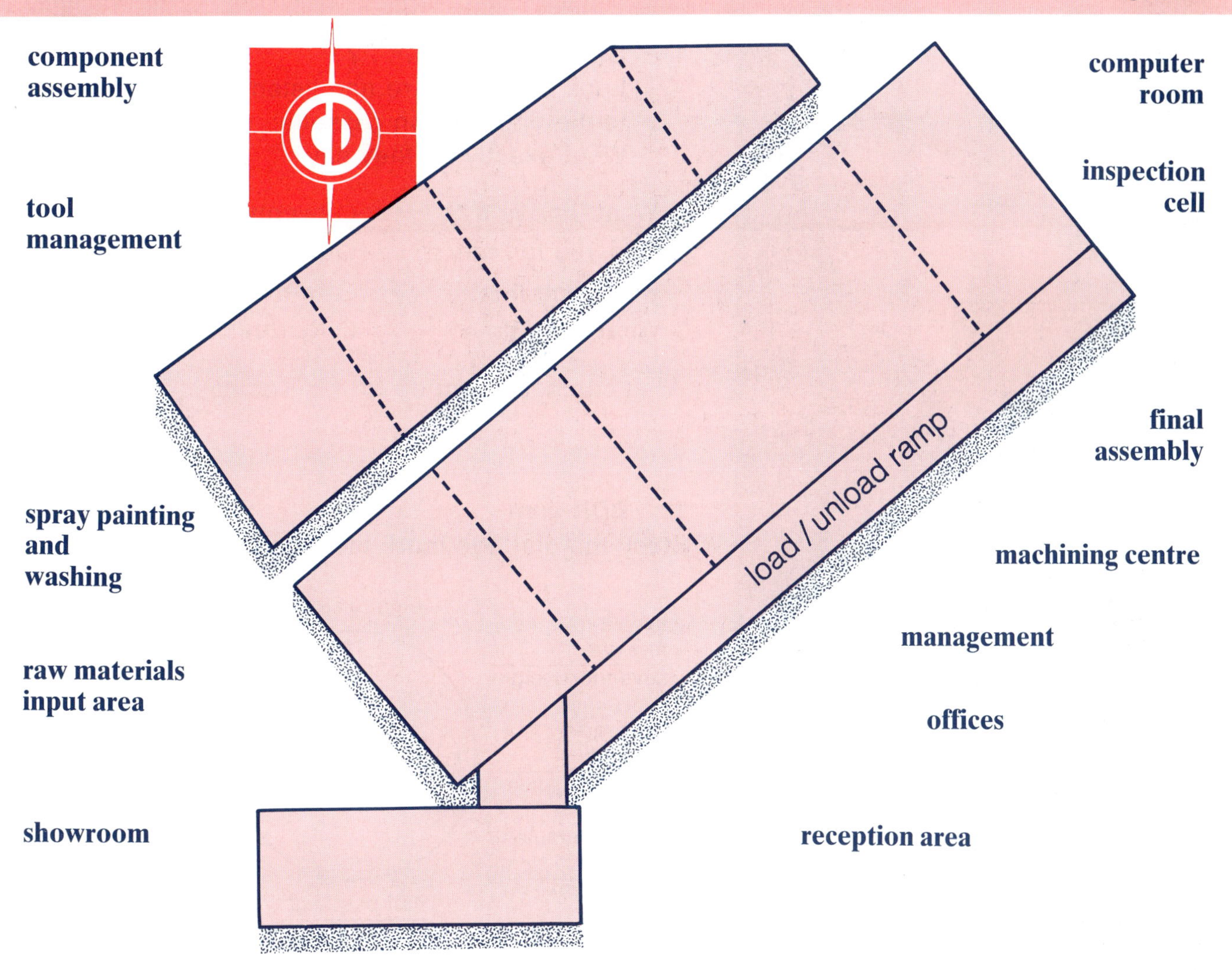

1 My company in Hove

Read Peter's letter to Martin.
Can you find C. Dugard Ltd. on the map?
What can you say about the company?

Where is it located?

....

What size is it?

....

What do they produce?

....

What can you say about the staff?
List all the jobs Peter mentions.

....

....

....

2 Plant layout

Listen to Peter's cassette letter and look at the plant layout of C. Dugard Ltd. Number the buildings in the drawing.

Listen again and connect the names of the areas or sections to the plant layout. Work with a partner.

3 Production flowline

Listen to Peter once more. Number the production areas or sections in the sequence they are mentioned.

4 Dear Peter

Write a short letter to Peter or record a cassette for him.
Tell him about your company, what they produce and which area you work in.

What time is it?

1 Can you tell me . . . ?

Look at the picture and listen to the dialogue on cassette: What exactly do they say?

- ○ What time is it?
- ○ Can you tell me what time it is?
- ○ What's the time?
- ○ twenty to three
- ○ two forty
- ○ twenty minutes to three
- ○ fourteen forty

What can/could **you** do for Mr Keaton?

....

2 BDR time

Study the dial and mark the times.

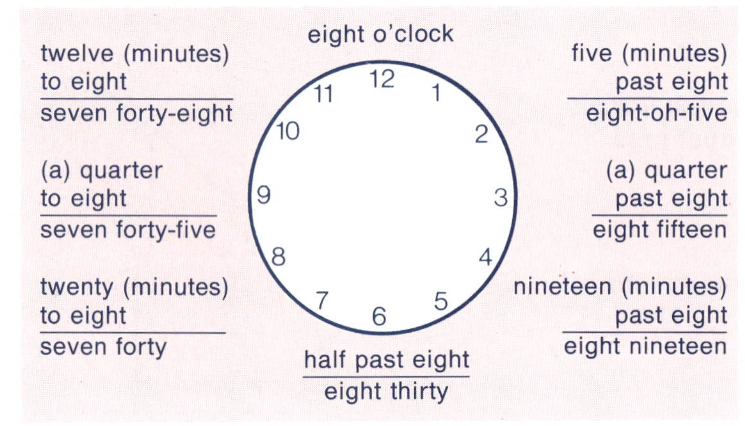

Listen to parts of the Blue Danube Radio morning programme. Can you mark the minutes?

3 More today at twelve noon

Listen again and take notes: What happens (or will happen) at what time?

Earth, Wind and Fire ....

The phrase of the day ....

Kissing was made illegal ....

But now the time: ....

The next ORF news ....

The next news on BDR ....

Time is told in different ways:
How do they tell it on BDR?

- ○ eight-oh-five
- ○ five past eight
- ○ five minutes past eight

It's
- ○ seven fifty.
- ○ ten to eight.
- ○ ten minutes to eight.
- ○ ten minutes before eight.

- ○ twelve o'clock
- ○ twelve noon

© Wayland

4 A technician's day at work

Look at the photos and read the text.
It's about Peter's day at work.
Tick the situations shown in the photos.

- ○ 8.15 a.m. clock in at work and change into working clothes.
- ○ 8.30 a.m. check the jobs to be completed.
- ○ 8.45 a.m. prepare tools, parts and supplies needed.
- ○ 9.00 a.m. start work on the job assigned.
- ○ 12.00 noon eat lunch with a friend.
- ○ 1.00 p.m. return to work, check fuses.
- ○ 4.00 p.m. tidy up work bench and tool box.
- ○ 4.15 p.m. change clothes and leave the company.
- ○ 5.00 p.m. fun-ride his dirt bike.

How many hours a day does Peter have to work?

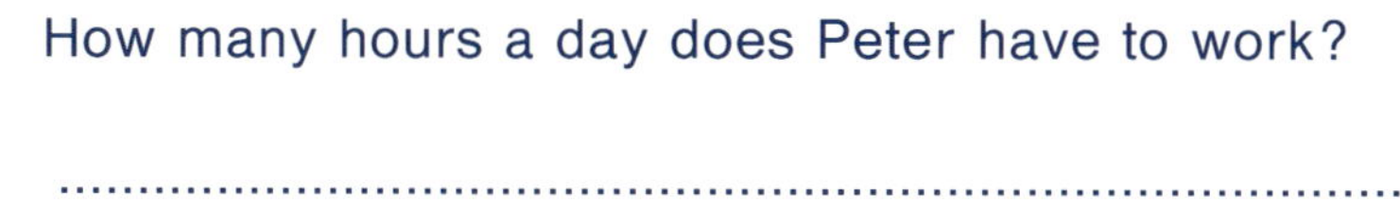

....

How long is his lunch break?

....

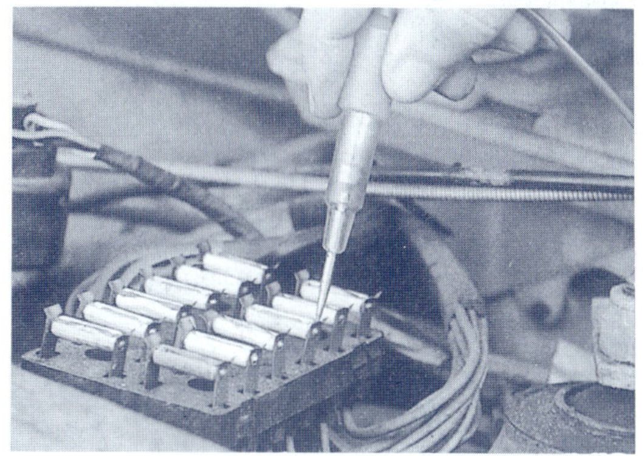
© Eaglemoss

5 A day in your life

When do you get up?
Do you cook breakfast? When?
When do you start work?
What do you do for lunch?
When do you get off work?
When do you eat supper?
What do you do in the evening?
What do you do on the weekend?

Tell your colleagues about it.
Or tell them about a routine that you would really enjoy.

Ask your partner about his/her working day and take notes into your copy-book.

When do you arrive at work?

When do you start your morning shift?

© KTM

EXTRACT OF INTERNATIONAL TIMETABLES
FAHRPLANAUSZUG
(WINTER 28 09 - 31 05)

dp ... DEPARTURE ar ... ARRIVAL

Berlin	dp	-	3.44	4.44	7.45	10.45
Braunschweig	ar	4.32	7.19	7.49	11.19	14.19
Hannover	ar	5.20	|	8.28	|	|
Köln/Cologne	ar	8.48	10.50	11.50	14.50	17.50
Dresden	dp	14.14			6.14	8.14
Frankfurt/M	ar	20.51			12.51	14.51
Köln/Cologne	ar	23.05			15.05	17.05
Karlsruhe	dp	4.45	7.59	8.59	11.59	14.59
Heidelberg	ar	5.45	|	|	|	|
Köln/Cologne	ar	8.45	10.59	11.59	14.59	17.59
München/Munich	dp	1.57	5.44	6.44	9.44	12.44
Mannheim	ar	|	8.32	9.32	12.32	15.32
Köln/Cologne	ar	8.33	10.59	11.59	14.59	17.59
Regensburg	dp	3.01	4.44		9.03	11.03
Nuremberg	ar	|	6.17		10.18	12.18
Frankfurt/M	ar	6.51	|		|	|
Köln/Cologne	ar	9.05	11.05		15.05	17.05
Köln/Cologne	dp	9.14	11.17	12.17	15.17	18.17
Aachen	ar	9.53	12.00	13.00	16.00	19.00
Lüttich/Liege	ar	10.35	12.42	13.42	16.42	19.42
Bruxelles	ar	11.42	13.55	14.55	17.55	20.55
Ostende	ar	12.48	15.09	16.09	19.09	22.09
London Vic	ar	16.07j	18.33j	19.33j	22.07j	6.30j
	ar	19.03				7.20

j) Jetfoil: special supplement required

Ihre Reiseverbindungen

Österreichische Bundesbahnen **Your travel-connections · Vos connections · I vostri colle**

Bahnhof / station gare / stazione		Uhr/hours heures ore	Uhr/hours heures ore	Uhr/hours heures ore	Uhr/hours heures ore
	ab/dp/pt				
	an/ar				
	ab/dp/pt				
	an/ar				
	ab/dp/pt				
	an/ar				
	ab/dp/pt				
	an/ar				
	ab/dp/pt				
	an/ar				

Auskunft ohne Gewähr · information without guarantee · information sous toutes réserves · informaz
W = an Werktagen · on weekdays · les jours ouvrables · nei giorni feriali
S = an Sonn- und Feiertagen · on sundays and holidays · les dimanches et jours fériés · si effettuano nei
X = umsteigen · change · changer de train · cambiare treno

1 When will I have to leave?

Listen to Marijana at the railway station.

Where does she want to go? ....

Where does she want to leave from? ....

Will she have to change?

Where? ....

Which train will she take? Listen again and mark it on the **Extract of International Timetables.**

Listen once more and fill in the **Travel Connections** form for Marijana.

2 am or pm?

Which do you think is right?

am ○ ○ 12.00 to 24.00

pm ○ ○ 00.00 to 12.00

Connect the boxes.

3 at, in or to?

Listen to the scene at the railway station and fill in the missing words.

Let's see
Here: There's one that'll get you ...... London Victoria ...... 6.30 ...... the morning.
Munich ...... departure 12.44 ...... the ...... of July.
Will I have to change?
Yes, ... Cologne. You arrive there ...... 17.59.

When do you use **at, in** or **to**? Formulate a rule in your own words.

Imagine this: you board a train at your local station in Germany and your vacation has already begun. From this moment on, you just have to relax and look forward to your stay in Great Britain. Read a book or linger in the dining car. Or watch the German countryside slip by outside your window.

Soon the green pastures and windmills let you know you've reached Holland. From here it's just a three-hour trip by train to the port of Hoek van Holland, where you get out right on the dockside. And then you can treat yourself like royalty. It's not so expensive – not since the "MS Queen Beatrix" has been around. It is the biggest and most modern ferry ever to sail between Holland and England. In seven short hours this luxuriously-equipped liner will take you to the port of Harwich. In the meantime, you can do some duty-free shopping in the well-stocked "North Sea Shopping Center" or enjoy a bite to eat in the cafeteria, open round-the-clock. There are self-service and à la carte restaurants on-board, too. Not to mention the stylishly decorated bars and lounges available for after-dinner drinks.

Of course, you can also cross the Channel via Oostende and Dover by ferry or with the speedy Jetfoil which takes only 100 minutes.

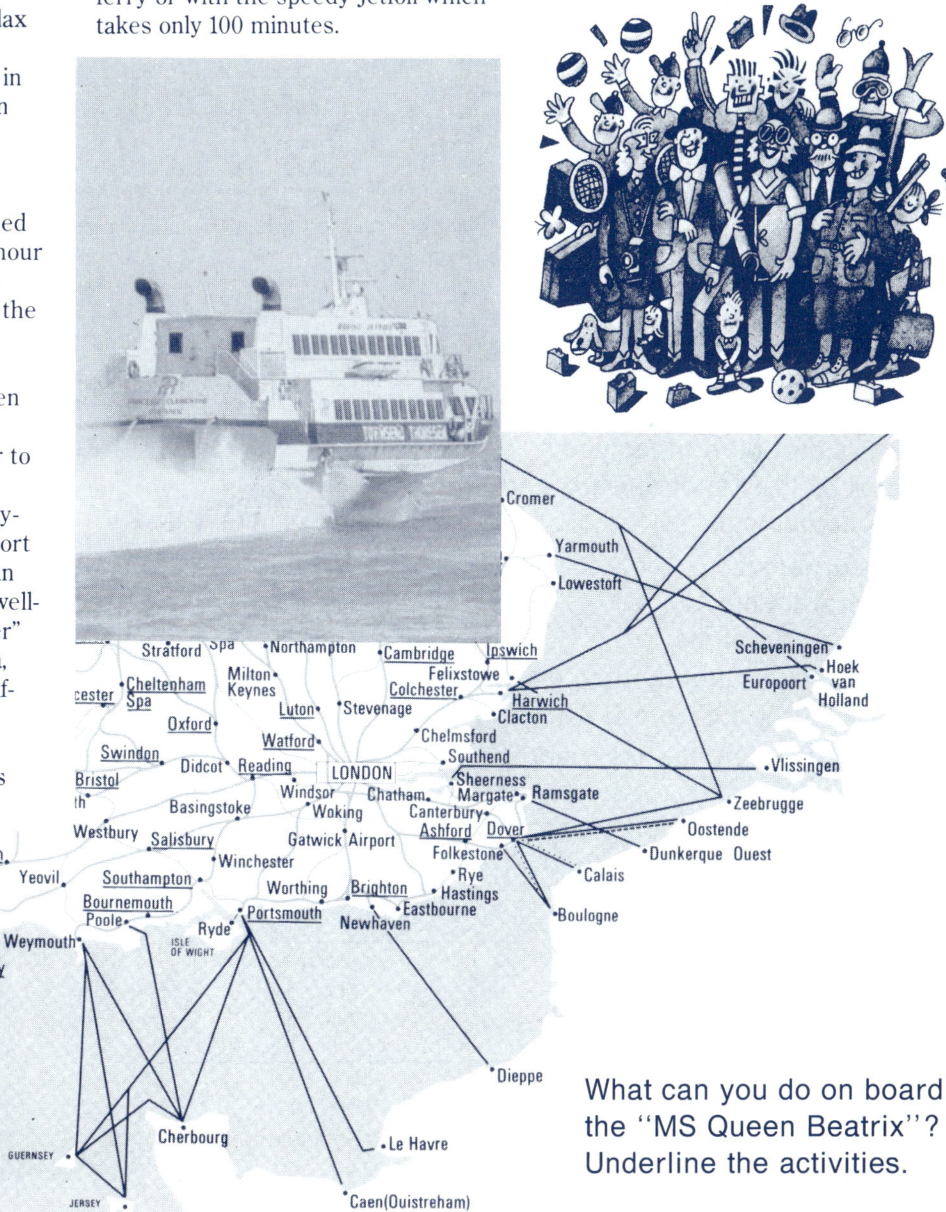

4 Across the Channel

On your way to England you'll cross the Channel. The article tells you more about it.

How long does the ferry crossing take?

From ....

to ....

And how long does the jetfoil take?

From ....

to ....

Mark the Channel crossings on the map.

What can you do on board the "MS Queen Beatrix"? Underline the activities.

LANDING CARD
Immigration Act 1971

Please complete clearly in BLOCK CAPITALS
Veuillez remplir lisiblement en LETTRES MAJUSCULES
Bitte deutlich in DRUCKSCHRIFT ausfüllen

Family name / Nom de famille / Familienname ....

Forenames / Prénoms / Vornamen .... **Sex** / Sexe / Geschlecht (M,F)

Date of birth / Date de naissance / Geburtsdatum: Day | Month | Year — **Place of birth** / Lieu de naissance / Geburtsort ....

Nationality / Nationalité / Staatsangehörigkeit .... **Occupation** / Profession / Beruf ....

Address in United Kingdom / Adresse en Royaume Uni / Adresse im Vereinigten Königreich ....

Signature / Signature / Unterschrift .... MW 574 330

For offical use / *Reserve usage officiel* / *Nur fur den Dienstgebrauch*

CAT | -16 | CODE | NAT | POL

1 The Landing Card

Listen to the scene on cassette. Where does it take place?

....

Who are the people talking?

....

Where are they going?

....

Who has to fill in a Landing Card?

- ○ Travellers from EU member states.
- ○ Travellers from non-EU countries.
- ○ Students from Germany.

EU ... European Union

LANDING CARD
Immigration Act 1971

Please complete clearly in BLOCK CAPITALS
Veuillez remplir lisiblement en LETTRES MAJUSCULES
Bitte deutlich in DRUCKSCHRIFT ausfüllen

Family name / Nom de famille / Familienname ....

Forenames / Prénoms / Vornamen .... **Sex** / Sexe / Geschlecht (M,F)

Date of birth / Date de naissance / Geburtsdatum: Day | Month | Year — **Place of birth** / Lieu de naissance / Geburtsort ....

Nationality / Nationalité / Staatsangehörigkeit .... **Occupation** / Profession / Beruf ....

Address in United Kingdom / Adresse en Royaume Uni / Adresse im Vereinigten Königreich ....

Signature / Signature / Unterschrift .... MW 574 333

For offical use / *Reserve usage officiel* / *Nur fur den Dienstgebrauch*

CAT | -16 | CODE | NAT | POL

2 Can I fill it in for you?

Look at the Landing Card and find the equivalent

for **surname:** ....
for **first name:**
and **middle name:** ....

Look at the Landing Card and fill it in for Tom D'Agostino.

Then complete the following part of the dialogue:

And where in England .... to stay?

At .... in Brighton, the Madeira.

Can you fill in a Landing Card for a traveller who has hurt his or her hand?

3 Can I see your passport, please?

Look at the pictures and listen:
Which situation comes first?
And which is second?
Number the pictures.

.... intend to stay .... Ms Lindner?

IMMIGRATION

Is this on business ....?

....

Have a pleasant stay ....

And what's ....

NOTHING TO DECLARE

Listen again and complete what the people say.

All right, ...

4 At the border

You can now go through immigration and customs.

Prepare your arrival in England in a group of three: one immigration officer, one customs officer, and a tourist. Then present your scene to the class. Which presentation did you like best? And the others?

FOR PASSENGERS OF 17 YEARS AND OVER WHEN PURCHASED ON BOARD.
Prices may be subject to change according to the availability of products.

INTO THE UK, HOLLAND, REPUBLIC OF IRELAND OR CHANNEL ISLANDS.		
WINES AND SPIRITS		
SPIRITS	Over 22% volume e.g. Whisky, Gin, Rum, Cognac	**1 LITRE**
OR	Fortified or Sparkling Wine e.g. Sherry, Port, Champagne	**2 LITRES**
OR	Aperitif Wine e.g. Cinzano, Martini	**2 LITRES**
PLUS	Still Table Wines	**2 LITRES**
TOBACCO PRODUCTS		
	Cigarettes	**200**
OR	Cigarillos (*Maximum weight each of 3g*)	**100**
OR	Cigars	**50**
OR	Tobacco	**250grammes**
PERFUMES AND TOILET WATERS		
	Perfume (*approx 2fl.oz. or 50g*)	**60ml**
PLUS	Eau de Toilette/After Shave	**250ml**
OTHER ARTICLES	To the value of	**£28**
IRELAND ONLY	(*IR£16 per person for persons under 15 years*)	**IR£32**

Would you like to register?

384 HOTELS & INNS

Iron Duke The, 3 Waterloo St, Hove .... **Brighton** 732724
Iverna Hotel—
32 Marine Pde,
Reception .... **Eastbourne** 30768
Visitors .... **Eastbourne** 21953
Jolly Sportsman Inn The, East Chiltington .... **Plumpton** 890400
Jolly Tanners The, Staplefield .... **Handcross** 400335

KINGS HEAD HOTEL—
Open all year to Businessmen Families Tourists,
South St, Cuckfield .... **Haywards Hth** 454006

THE KINGS HOTEL
139/141 KINGS ROAD, BRIGHTON
SEAFRONT HOTEL–95 BEDFOOMS
VIRTUALLY OPPOSITE WEST PIER
BALLROOM–BAR–LIFT–SUN LOUNGE
TV LOUNGES–GAMES ROOM
LARGE RESTAURANT
CATERING FOR 175 GUESTS
PRIVATE PARTIES–WEDDING
RECEPTION, DINNER/DANCE
RESIDENT OR NON RESIDENT CONFERENCES
FOR UP TO 120 DELEGATES
BRIGHTON 29133

Kings Hotel, The, 139 Kings Rd .... **Brighton** 29133
Kingsway Hotel, 2 St. Aubyns, Hove .... **Brighton** 722068
Kingsway Hotel, 2 St. Aubyns, Hove .... **Brighton** 732575
Ladies Mile Hotel, Mackie Av .... **Brighton** 554647
Lamb Inn, Lambs Gn .... **Rusper** 336
Langfords Hotel, 12 Third Av, Hove .... **Brighton** 738222
Madeira Hotel, 19 Marine Pde .... **Brighton** 698331
Madeira Hotel, 21 Marine Pde .... **Brighton** 607853

MAGNOLIA HOUSE—
Small Select Hotel—Modern Comforts—Good Prices,
274 Dyke Rd, Brighton, E. Sussex .... **Brighton** 552144

Majestic Hotel
26 Royal Pde, Reception .... **Eastbourne** 30311

Metropole Hotel—
Kings Rd .... **Brighton** 775432
(Telex 877245)

Michel E.E.C, Sunnyside Hotel 5 Portland Pl .... **Brighton** 601988

1 Accommodation

Look through the hotels in the **Yellow Pages** of the Brighton telephone directory. Which of the hotel names do you find most remarkable? Where would you want to stay?

"Have a nice day."

2 Confirming a reservation

Listen to Tom:

Where is he? ....

Which hotel is he calling? ....
What sort of room did he book?

....

Fill in the BOOKING FORM for Tom.

What sort of room would **you** book?
At which hotel?
And your partner?
Reserve a room for him or her.
Design your own booking form.

YOUR ALL-SEASONS ACCOMMODATION BOOKING FORM

Name of hotel or guest house you wish to book:

Hotel/guest house ....

Name

Telephone number:

Approximate arrival time: AM PM

GIVE THIS FORM TO YOUR TRAVEL AGENT OR SEND DIRECT TO THE HOTEL WITH WHICH YOU HAVE MADE YOUR PROVISIONAL BOOKING

NAMES				BEDROOMS							MEALS			DATES			COST
Surnames of proposed hotel guests (IN BLOCK CAPITALS)	Initials	Title Mr Mrs Miss	Age if under 18	Twin room	Double room	Single room	Family room	Cot	Private shower	Private wc & bath/ shower	B* and B	Half board	Full board	Arrival date	Depart-ure date	No. of nights	£

If dinner is included in the arrangement you are making do you require an evening meal on your day of arrival? ☐ YES ☐ NO

TOTAL (£)

* Bed and Breakfast.

I have read and agree to abide by the conditions of booking and enclose my crossed cheque/postal order to the value of:

£ .... as Deposit of £10.00 per person/Full payment* *(please delete as applicable)* made payable to:
Name of hotel ....

Date ....

Signed ....
**Full payment is required when arrival is within six weeks of booking date*

3 Here's your key

Design a guest register in your copy book like this.

Date	Room No.	Name	Home address	Signature

Tom is registering at the Madeira Hotel.
Fill in the guest register for him.

Complete the following statements.

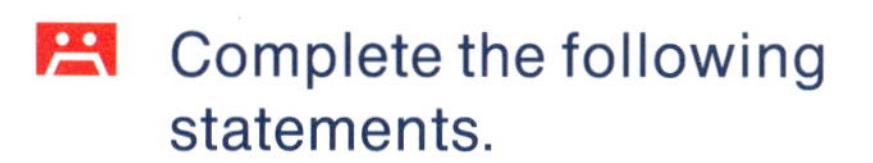

4 Let me do it for you

Form groups of four.
You can be the **receptionist** at the Madeira or another hotel.
Help your **guests** to register and fill in a GUEST REGISTER or a REGISTRATION FORM for them.

Maybe you want to practise spelling your name and address again?
When you are ready present your scene to the others.
Listen to the others and take notes.

5 A conference in Brighton

Look at the list of hotels in the Yellow Pages of the Brighton telephone directory.

Which hotel offers facilities for a conference?

For how many participants?

The Kings Hotel provided the checklist for conference facilities on the right.
Which of these facilities would you or your company need for a sales conference?
Make notes and compare them with a partner.
Then write a short letter to The Kings Hotel and reserve rooms and conference facilities for, say, 25 participants.

The RESTAURANT Revue

WHERE-TO-EAT GUIDE TO RESTAURANTS IN SUSSEX

Restaurant Cafe
An excellent selection of Creole, and French American cuisine presented in a way you will not have seen before
Open from 11 a.m. till 3 p.m. and 7 p.m. until you finish eating.
6 Little East Street, Brighton
Telephone (0273) 735527

69 St. James's Street, Bookings Btn. 674472
OPEN 7 EVENINGS

WOODIES WINE BAR
Open 7 days, lunch and dinner. Good food. Specialising in seafood. Informal atmosphere. Business lunches. Resident guitarist some evenings. Res. 736510. 93 CHURCH RD. HOVE

164698/17
THE FRIAR TUCK
18 York Place, Brighton
Tel: Brighton 601530
Fish and chip restaurant and take-away service ★ Special offer for OAP's ★ Fish and Chips at unbeatable prices ★ Special offer every Monday ★ Childrens portions at half price
OPEN 7 DAYS A WEEK
11.30 a.m.-midnight on Sunday, Monday, Tuesday and Wednesday
11.30 a.m.-3 a.m. on Thursday, Friday and Saturday

THE PANDA CHINESE RESTAURANT

4-COURSE DINNER MENU
£5.50
55 MARINE DRIVE ROTTINGDEAN. Tel. 35894
CLOSED ALL DAY TUESDAY

BLACK LION HOTEL
London Road, Patcham
Tel. Brighton (0273) 501220 or 501656
SOCIETY RESTAURANT
OPEN DAILY
Lunch 12-2 p.m. (except (Saturdays)
Full a la carte and table d'hote. Eves. (except Sundays) 7 p.m. till late
Full a la carte
Dinner Dance Saturday

TIFFIN:
TUES-SAT 12.30-2.30
DINNER:
MON-SAT 6.30-11.00

60 Western Road · Hove
Telephone No. (0273) 21343

THE FRENCH CONNECTION
Mon.-Sat. Lunch and Dinner
FULL LUNCH INC. COFFEE £8.50
4-COURSE DINNER £10.65.
Full a la Carte available
Private and Business Parties catered for
11 Little East Street. Tel. 24454

ROMANO RISTORANTE ITALIANO
Tuscany cuisine. Now open 7 days a week for lunch and dinner for you to try his exciting new menu. For reservations tel. Brighton 23574. 44 Preston St. (next to Chequers pub).

CHRISTY'S
The Italian restaurant with a difference Live music 5 nights a week. Open 7 days. Licensed till 2 a.m. 44 Kings Road between Middle Street and Ship Street. Telephone Brighton 25229

SEAFOOD £7.95
Special 5-course menu, available Monday to Saturday
Full a la carte also available
BUTTIMER'S RESTAURANT
Newhaven. Tel. 514632
The best selection of seafood you can wish for — oysters, mussels, crab, lobster.
Spanish and French cuisine.
Also best Scotch steaks
Lunches Tues.-Fri.; Dinners Mon.-Sat.

I like vegetarian food.

All right. Let's go to the Greenhouse.

1 Where shall we go?

Look through the Sussex Restaurant Guide.
What sort of restaurants can you find?

- ○ traditional (English)
- ○ elegant
- ○ vegetarian
- ○ seafood
- ○ informal
- ○ fish and chips
- ○ French
- ○ Italian
- ○ Mongolian
- ○ Indian
- ○ Chinese
- ○ French American

Which restaurant is the most expensive?
What do you think?

....

And the least expensive?

....

"The flowers look good!"

2 What sort of food do you like?

Listen to our friends (on cassette):
Look at the Sussex Restaurant Guide and tick the restaurants that are mentioned.

Which one did they choose?

....

What sort of food can you get there?

....

What sort of food do **you** like?

....

And your neighbour?

....

Where would **you** like to go?

....

3–4 Duke St.
Brighton
Brighton 23501

5–9 Woodstock Rd.
Oxford
Oxford 511995

SPAGHETTI

All spaghetti orders served with Garlic Bread and Mixed Salad

ITALIAN TOMATO SAUCE *onion, peppers, tomato and red wine* **£3.45**
TRADITIONAL MEAT SAUCE *prime mince, herbs and wine* **£3.45**
BRIGHTON SEAFOOD SAUCE *mussels, prawns and more* **£3.45**
A PORTION FOR CHILDREN *without a salad and garlic bread* **£2.25**

SALADS

With a choice of Blue Cheese, Thousand Island, French Dressing, Mayonnaise, Garlic Dressing

MRS BROWNS VEGETARIAN **£4.25**
TUNA FISH SALAD **£4.35**
AVOCADO, BACON AND SPINACH SALAD **£4.25**
CHEF'S CHICKEN AND HAM **£4.45**

MEAT FISH AND SPECIAL

All dishes served with Fried or Baked Potato (with butter or sour cream) and Mixed Salad, or Vegetable of the Day

SCOTCH SIRLOIN STEAK *half pound, with herb butter* **£6.85**
BROWNS LEG OF LAMB *chargrilled with rosemary served with Oxford Sauce* **£6.25**
PRIME ROAST RIBS *with barbecue sauce* **£4.85**
FISHERMANS PIE *with Cheddar Cheese Crust* **£4.25**
COUNTRY CHICKEN PIE **£4.25**
CHARGRILLED TURKEY BREAST **£4.65**
FRESH VEGETABLES WITH HOLLANDAISE SAUCE **£4.25**

BROWNS HAMBURGER
6oz of prime ground beef in a sesame seed bun
plain **£2.95** with melted cheese **£3.15** with bacon and melted cheese **£3.35**

SIDE ORDERS

Fresh Mushrooms **£1.15**
Baked Potato **95p**
Mixed Salad **95p**
Fried Potatoes **95p**
Garlic Bread per Person **60p**
Fresh Vegetable of the Day **95p**

DESSERTS

Chocolate Cake **£1.20**
Cheesecake **£1.20**
ICE CREAM £1.05
Chocolate, Strawberry, Vanilla or Coffee

Banana Cream Pie **£1.20**
Chocolate Pie **£1.20**
Apple Pie **£1.20**

SOFT DRINKS

PURE FRUIT JUICES
Orange, Apple, Grapefruit, Pineapple, Tomato **65p**

ICED DRINKS
Coke **65p**
Lemonade **60p**
Milk **60p**

TEA, COFFEE

POT OF TEA PER PERSON **55p**
Darjeeling
Ceylon, Jasmine
Earl Grey

REGULAR *black or white* **50p**
DOUBLE ESPRESSO **55p**
CAPUCCINO **55p**
ESPRESSO **50p**
HOT CHOCOLATE **55p**

Cheques only accepted with Bankers Card
Credit Cards not accepted
Value Added Tax 17,5% is included in the price

A service charge of 10% will be added to parties of 5 or more

Opening times, 11am–11.30pm Monday to Saturday,
12am–11.30pm Sundays, Bank Holidays

3 Here we are at Browns

You are invited for a four course dinner, on one condition:
You can only choose dishes you know.
Don't use a dictionary!
And don't worry about prices.
It's all paid for.

What will you have ...

... as a **starter?** ....

... as a **main course?** ....

... as a **side order?** ....

... for **dessert?** ....

And you can choose a **soft drink.**

....

Are you interested in any other dishes?
Now you can use a dictionary ...

4 Are you ready to order?

Listen to our friends.
Where do you think they are?

....

Now listen and look at the menu.
Tick the dishes they order.

5 I'll try the Crab and Avocado Salad

Now you order the meal you chose.
Are you ready?

Are you ready to order?
Just about. .... Mrs Browns Vegetarian Salad.

Form groups of 3 or 4.
One of you can be the **waiter** or the **waitress.**

AUG.07 04:59PM

1368 GERRY 17

1368 PREV BAL		.00
1 BLACK COFFEE	.50	
1 ESPRESSO	.50	
1 CHOC CAKE	1.20	
2464 BALANCE		2.20
1368 PREV BAL		2.20
1 WHITE COFFEE	.50	
2477 BALANCE		2.70
1368 PREV BAL		2.70
2504 CASH		2.70

Browns
Oxford 511995
Brighton 23501

Service is not included
except for parties of 5 or more

TOTAL £ 2=70

Please pay at your table

VAT No. 410818185

1 Here's your bill . . .

Look at the menu on page 27.
Do you remember the meal our friends ordered?
How much will it cost?
Can you guess?

Look at the bill and listen to what's happening at Browns.
What's wrong with the bill?

2 How much will it be?

Can you work out the bill with a partner?
Look at the menu on page 21 and tell your partner what they had:

There's the Vegetarian Salad, it's four pounds twenty-five.

Four twenty-five, yes

Then there's ....

Your partner takes notes on the bill.
How much will it be?

3 The bill please!

Remember the meal **you** ordered?
Ask for the bill.
Your partner can be the waiter or waitress.
Check the bill against the menu.

Don't forget to **add 12 to 15% for service.**
Then hand it to the one who invited you.

"Do we have to do this every time to decide who pays the bill?"

4 What about desserts?

Listen to the discussion
at Browns again,
to the part about the desserts.
Look at the cover of the menu.

Which words do you understand?
And your neighbour(s)?

....

What can you say about the Valencia?

....

....

A.	Selection of Danish Pastries	.90
B.	French Apple Pie	.90
C.	Black Forest and Kirsch Gateaux	1.20
D.	American Cheesecakes Assorted	1.00
E.	Coffee, Mandarin and Tia Maria Gateaux	1.20
	Ice Cream Sundae	
F.	Selection of three flavours of Ice Cream	1.10
G.	Calypso Banana Boat: Sliced banana, coconut ice cream, strawberry sauce, topped with chocolate flake, cherries and whipping cream	1.80
H.	Tiki Fruit Cup: Orange ice cream on fruit salad and whipping cream	1.75
	ICE CREAM COCKTAILS	
J.	Chocolate Mint Fizz Bacardi, chocolate ice cream and Creme de Menthe	2.50
K.	Frozen Alexandras Brandy, brown Cream de Cacao, vanilla ice cream	2.50
L.	Nassau Express Drambuie, Brandy, coffee ice cream	2.50
M.	Fruit Salad and Cream or Ice Cream	1.50

5 Sounds delicious

Go through the menu and underline all the words you understand; also the ones you can guess.

Would you like to order a dessert?
What would you say?
Work with a partner.
Present it to your class …

World money

1 Post office or bank?

Look at the picture:
Is this a post office?
Or a bank?
Or what?
Now read the text below and underline all the services they offer.

Do we need any more English money?

Post Office
Girobank

National Girobank has offices at almost every post office in Great Britain. They are open longer hours than other banks, and on Saturday mornings, too. And you can enjoy all the usual services you would expect from any other bank.
● Cheque book ● Cheque card ● Cash card
● Visa card ● Personal loans and credits
● Savings deposit accounts ● Foreign exchange

When in England what services would you require from a bank?

2 At the foreign exchange counter

Listen to what is happening inside the bank, at the foreign exchange counter.
Connect the boxes.

Dennis ○
Amy ○

- ○ wants to change German marks.
- ○ wants a Visa card.
- ○ wants to cash travellers cheques.
- ○ receives £ 50.
- ○ deposits 200 German marks.
- ○ receives £ 81.63.
- ○ pays DM 2.45 per pound sterling.
- ○ takes out a personal loan.
- ○ must show his or her passport.

THE £ TODAY

Australia	2.23
Austria	17.00
Belgium	50.50
Canada	1.92
Denmark	9.53
France	8.33
Germany	2.45
Greece	334.00
Hong Kong	11.55
India	44.42
Ireland	0.95
Italy	2,265
Malta	0.54
Netherlands	2.77
New Zealand	2.88
Norway	10.57
Portugal	220.00
Saudi	5.77
Spain	173.00
Sweden	11.27
Switzerland	2.24
Turkey	12.90
USA	1.52

Approximate tourist rates

Daily Mail

Pounds Sterling

Denominations
Stg£ 10
Stg£ 20
Stg£ 50
Stg£ 100
Stg£ 200

How do you take money with you when you go abroad?
What can you do when you lose cash?
And what if you lose travellers cheques?

3 Mark values

Look at the foreign exchange rates for the American (US) dollar and for the British Pound (Sterling).

How many marks will you have to pay for £ 1 (one British Pound Sterling)? ....

And for US$1 (one US dollar)? ....

4 Foreign exchange rates

Listen to the exchange rates on Radio International and complete the foreign exchange chart.

Denom. *)	Country	Nationality	Currency	Value DM
	US			....
	UK			
	CDN			
	AUS			
	CH			
	B			
	F			
	NL			
	A			
	E			
	I			
	DK			
	S			
	N			
	JAP			

*) Denomination of foreign currency: 1, 100 or 1000

MONDAY'S FOREIGN EXCHANGE RATE:

	Foreign currency per dollar Mon.	Fri.
Australian dollar	1.488427	1.486989
Austrian schilling	11.49	11.59
Bahrain dinar	0.3768	0.3768
Belgian franc	33.6125	32.925
Brazilian cruzeiro	na	na
British pound	0.645369	0.655308
Brunei dollar	1.6355	1.6355
Canadian dollar	1.27395	1.28215
Chinese renminbi	5.7518	5.7518
Cypriot pound	0.468187	0.468187
Danish krone	6.3221	6.32
Dutch guilder	1.83605	1.85205
Egyptian pound	3.2659	3.2659
European Currency Unit	0.832605	0.841361
Finnish markka	5.445	5.485
French franc	5.5545	5.6065
German mark	1.63315	1.6485
Greek drachma	218.4	220.15
Hong Kong dollar	7.742	7.743
Indian rupee	28.7	28.7
Indonesian rupiah	2049.0	2049.0
Irish punt	0.602864	0.602864
Israel shekel	2.7328	2.7328
Italian lire	1494.75	1520.25
Japanese yen	125.33	125.455
Jordanian dinar	0.6752	0.6752
Kenyan shilling	35.56	35.56
Kuwaiti dinar	0.2849	0.2849
Lebanese pound	1838.0	1838.0
Malaysian ringgit	2.5936	2.5975
Mexican peso	3.105	3.114
Moroccan dirham	8.778	8.778
New Zealand dollar	1.954079	1.95944
Norwegian krone	6.9795	7.0535
Oman riyal	0.385	0.385
Pakistani rupee	25.59	25.59
Philippine peso	24.61	24.61
Portuguese escudo	146.6	148.0
Qatar riyal	3.64	3.64
Saudi Arabian riyal	3.7406	3.7406
Singapore dollar	1.6633	1.6633
South African rand	3.087	3.087
South Korean won	775.0	775.0
Spanish peseta	117.025	117.025
Sri Lankan rupee	44.86	44.86
Swedish krona	7.40335	7.40335
Swiss franc	1.5055	1.5055
Taiwanese dollar	25.16	25.16
Thai baht	25.41	25.41
Turkish lira	8645.0	8645.0
UAE dirham	3.6725	3.6725

USA TODAY/International Edition ·
Source: Riggs National Bank of Washington, D.C.

Have you noticed that there are several ways of saying the numbers?

How does the speaker do it?
Listen again and take notes.

5 Holiday money

Pick a country where you would like to spend a week or two.
Take enough money to the bank, say 1,000 German marks, and exchange it for the foreign currency you need.
Your partner works at the foreign exchange counter of a bank; he or she **does not** speak German.
Practise your scene and present it to your class.
Then listen to the others and take notes: name of customer, country, currency and the amount he or she gets.

1 I need stamps

Look at the photo and listen to the dialogues.

What does Amy, the first customer, need?

....

What for?

....

How many (of each)?

....

Royal Mail International Letters

All-up rates for Europe

There is only one class of mail available for letters to Europe, ie 'All-up'. The rates charged are the same as those for surface letters outside Europe, but the items are sent by air whenever this will result in earlier postal delivery.

Air mail labels or envelopes with coloured borders should **not** be used.

Letters and Postcards

Not over	£	p	Not over	£	p	Not over	£	p	Not over	£	p
20g		22	250g	1	06	500g	2	02	1750g	5	17
60g		37	300g	1	25	750g	2	77	2000g	5	72
100g		53	350g	1	44	1000g	3	52			
150g		70	400g	1	64	1250g	4	07			
200g		88	450g	1	83	1500g	4	62			

Issued by Royal Mail Letters.

When this leaflet was printed, all the information was correct. However, as things do change from time to time, you can always obtain up to date information from your local post office.

2 A letter to Germany

Now look at the postal rates for Europe and listen again.

How much is a letter to Germany?

....

And how much is a postcard?

....

How much extra does 'special delivery' cost?

....

3 'Definitive stamps' and others

The Royal Mail 'definitive stamps' show a portrait of the Queen in profile. The basic design is more than 150 years old, the present series has been in use since 1967, although different values were added whenever the rates changed.

What do you think of the classic definitive stamps?
And how do you like the traffic motives?
Would you like to buy some stamps at a post office in Britain?
Your partner can be a post office clerk.

Royal Mail Special Delivery

Factfile
FROM THE GUINNESS BOOK OF RECORDS

DOGS WERE EMPLOYED TO HAUL MAIL-CARTS IN THE COASTAL TOWNS OF SUSSEX IN THE LATE NINETEENTH CENTURY. THE SERVICE WAS TERMINATED FOLLOWING COMPLAINTS FROM ANIMAL LOVERS.

Drawn by DICK MILLINGTON

4 How it all works

Collections of mail are made several times a day and taken, usually by van, to a main posting office.

Here, POP letters are separated from packets and large envelopes, then turned the right way up, divided into first and second class and the stamps are cancelled.

Next, each postcode is translated into two rows of blue dots printed on the envelope. These dots can be "read" by automatic sorting equipment.

POP . . . Post Office Preferred size of envelopes, not smaller than 90 × 140 mm and not larger than 120 × 235 mm, can be handled by Post Office machines.

first class mail . . . standard letters and postcards.

second class mail . . . printed matter at lower rates within the UK.

Read the paragraph from a British Post Office brochure.
What do the post office workers do?
Tick the correct statements which are written in **active voice.**

Read the paragraph again.
What is done to the mail/letters?
Convert the statements on the left into **passive voice.**

○ They collect the mail.	*The mail is collected* ....
○ They separate the POP letters.	*The POP letters are separated* ....
○ They throw away the postcards.	....
○ They turn the letters face up.	....
○ They put on extra stamps.	....
○ They cancel the stamps.	....

5 Malcolm Biles

Read this paragraph from *The Kingdom By The Sea,* by Paul Theroux, an American who wrote about his journey around the coast of Great Britain in 1983.

What can you say about the paragraph?

What does it say about Malcolm Biles?

About the British post office?

And what about dog licences?

tick someone off . . . criticize someone

quid . . . slang for £

Malcolm Biles asked for a look at my map. He was twenty-three, a post office clerk from Inverness, who was on a cheap day-return. I had wanted to meet a post office worker, I told him. British post office workers did much more than sell stamps. They processed car licences, television licences, Family Allowances, pensions, Inland Telegram Postal Orders, all the tasks required by the Post Office Savings Bank, and a hundred other things. They had seven weeks' training and the rest had to be learned on the job, in full view of the impatient public. It was Malcolm who spoke of the impatience. People were much ruder than they used to be and some of them stood there and ticked you off!

'What about dog licences?' I asked.

Dog licences! It was Malcolm Biles's favourite subject. The price of a dog licence was $37^{1}/_{2}$ pence, because in 1880 it had been fixed at seven shillings and sixpence. The fee had never been changed. Wasn't that silly? I agreed it was. There were six million dogs in Britain, and only half of them were licenced. But the amazing thing was that it cost four pounds (almost seven dollars) to collect the dog licence – the time, paperwork, and so forth.

'Why not abolish the fee?' I asked.

Malcolm said, 'That would be giving up.'

'Why not increase it to something realistic – say, five quid?'

'That would be unpopular,' he said. 'No government would dare try it.'

'How long do you figure you'll be staying in the post office?'

'For the rest of my life, I hope,' he said. The train jolted. 'Ah, we're away.'

1 Big business

Look at the drawing:
Who creates . . . what?

....

And who cleans it up?

....

"We create it, we clean it up – business couldn't be better . . .!"

Pollution

Words and music by Tom Lehrer

If you visit American city
You will find it very pretty.
Just two things of which you must beware:
Don't drink the .... and don't breathe the .....
Pollution, pollution,
they've got ....
Turn on your tap,
and you get hot and cold running ....
See the halibuts and the ....
Being wiped out by ....
...... gotta swim and ...... gotta fly,
But they don't last long if they try!
Pollution, pollution:
You can use the latest toothpaste,
And then rinse your mouth with industrial ....
Just go out for a breath of ....
And you'll be ready for ....
The city streets are really quite a thrill:
If the hoods don't get you, the ....
Pollution, pollution: wear a gas mask and a veil:
Then you can breathe, 's long as you don't inhale.
Lots of things there that you can drink,
But stay away from the kitchen sink:
The breakfast .... that you throw into the bay
They drink at lunch in San José!
See the crazy people there,
Like lambs to the slaughter
They're drinking the ....
And breathing (gulp) ....

2 Pollution

Listen to this song by Tom Lehrer and look at the gap text.

Then look at the following words and make sure you know what they mean. Look them up in a dictionary if necessary.

smog ....
sewage ....
mud ....
sturgeons (a fish) ....
detergents ....
waste ....
monoxide ....
veil ....
garbage (US for rubbish) ....

Medicare ... medical care
crud .... dirt
hoods .... criminals
the bay ...... San Francisco Bay
San José ... city at the southern end of the bay

How would the words fit into the gaps?

Listen to the song again and fill in the words.

Now that you understand the lyrics what do you think of the song?
By the way, **Pollution** was recorded in 1965, at the **Hungry I** club in San Francisco.

"We've been coming here for our picnics for years."

3 Environmental Protection Quiz

What can **you** do to protect your environment?

Read the following advice and mark each of them, like this:

3 points :	Yes, good idea. I already do that.
2 points :	Good idea. I'll do that.
1 Point :	Sounds okay. I should try that.
No point :	Looks like too much trouble.

- ○ Buy milk and soft drinks in glass-bottles which are re-usable.
- ○ Don't accept bottles and cans which cannot be returned.
- ○ Don't buy things wrapped in plastic.
- ○ Collect re-usable materials, such as paper, clothes, glass and waste metals. They can be recycled.
- ○ Collect chemicals, such as batteries, medicines etc. and turn them in at special deposits.
- ○ Ride a bike or walk or travel by public transport whenever possible instead of using a car or motorbike.
- ○ Save water – take a shower instead of a bath.
- ○ Save energy – electricity, gas, oil. Switch off the lights when you leave a room.
- ○ Put on a pullover when you feel cold. Don't turn up the heat.
- ○ Keep informed about environmental issues.
- ○ Join an environmental protection group.
- ○ Don't throw things away without asking – maybe someone else could use them!

Add up your score and write it down here **before** looking at the key for evaluation (at the bottom of the page).

Are you doing enough for your environment?

Evaluation

32–36	Very good. Continue the good work.
27–31	Good. But you can do better!
22–26	Okay. But try to do better.
17–21	Are you doing enough for your environment?
0–16	Sorry! You seem to be part of nature's problem ...

4 Protection from above

What do you think the government should do about environmental protection?

I think they should ...

... build railroads.

... reward the saving of energy.

... restrict private transport.

... keep people informed about environmental issues.

And they shouldn't ...

A mechanic's tool kit

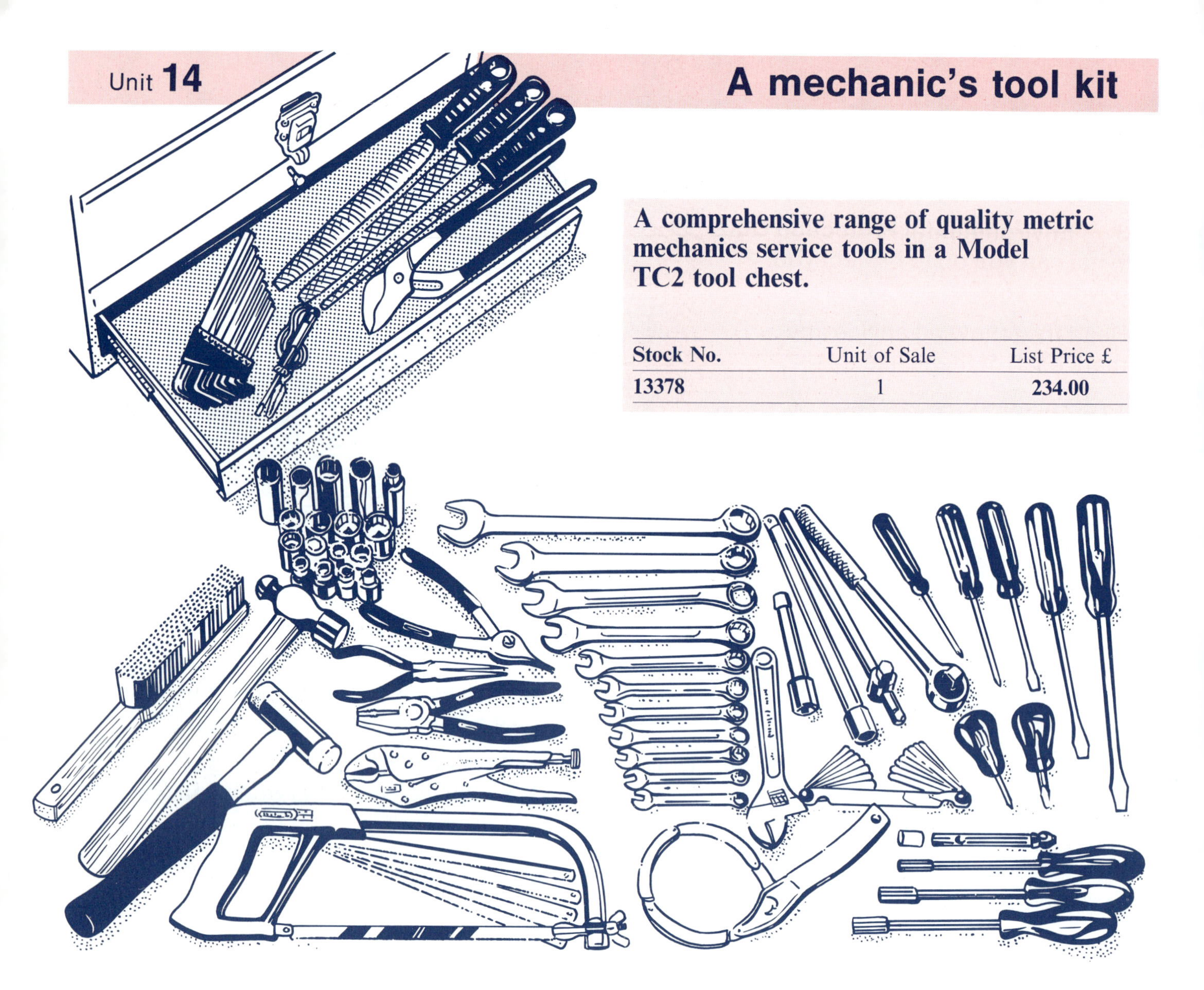

A comprehensive range of quality metric mechanics service tools in a Model TC2 tool chest.

Stock No.	Unit of Sale	List Price £
13378	1	**234.00**

Contents

$\frac{1}{2}$" sq. drive sockets and accessories

1 reversible ratchet
2 2 extension bars size: 125 mm (5") and 250 mm (10")
3-11 bi-hexagon sockets size: 8, 9, 10, 11, 12, 13, 14, 17, 19, 22 and 24 mm
4 4 bi-hexagon deep sockets size: 10, 13, 17 and 19 mm
5 sliding tee bar
6 2 spark plug sockets size 10 and 14 mm

7 11 combination spanners size: 6, 7, 8, 9, 10, 11, 12, 13, 14, 17 and 19 mm
8 adjustable wrench 200 mm (8")
9 self-grip pliers 280 mm (10")
10 waterpump pliers 240 mm ($9\frac{1}{2}$")
11 reversible circlip pliers 200 mm (8")
12 combination pliers 200 mm (8")
13 radio pliers 160 mm ($6\frac{1}{4}$")

14 second cut handled file–flat–200 mm (8")
15 second cut handled file–round–200 mm (8")
16 second cut handled file–half round–200 mm (8")
17 ball pein hammer 454 gms (1 lb)
18 soft faced hammer 454 gms (1 lb)
19 3 plain slot screwdrivers size: 100 mm (4"), 150 mm (6") and 200 mm (8")
20 plains slot chubby screwdrivers 35 mm ($1\frac{3}{8}$")
21 2 cross slot screwdrivers No. 1 and 2
22 cross slot chubby screwdriver No. 2
23 3 nut spinners size 5, 6 and 8 mm
24 feeler gauge MM/imperial
25 car electrics tester
26 tyre pressure gauge
27 oil filter wrench
28 set of ball point hexagon keys size: 2·5, 3·0, 4·0, 5·0, 6·0, 8·0 and 10·0 mm
29 adjustable hacksaw frame 300 mm (12") with five spare HSS blades
30 wire brush

1 Service tools

Look at the tools on the opposite page and at the list of contents below.
What are the tools called in English? Number each tool in the drawing.
Compare your results in a group of 3 or 4.
Study the following examples and use similar questions and answers.

Number 9 are self-grip pliers.
What about number 11?
Those are circlip pliers.
To fit and remove circlips?
Right.

circlips

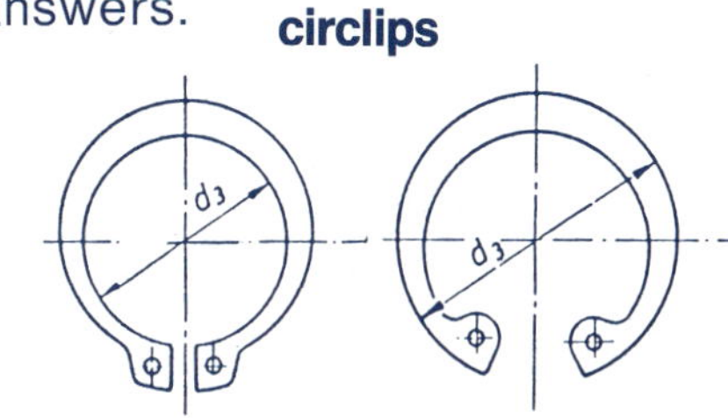

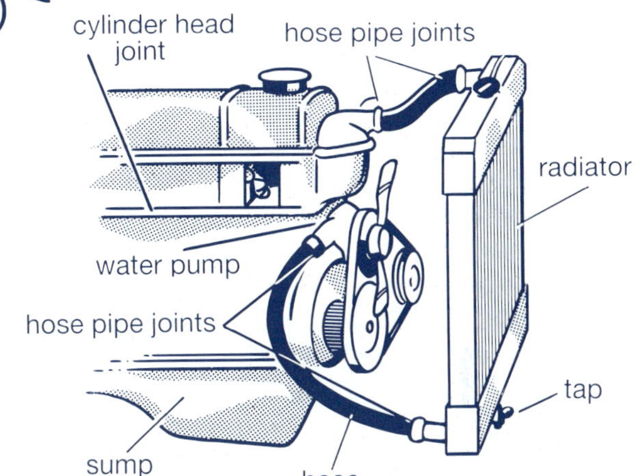

2 A new radiator

Look at the list of tools on the opposite page and listen to the cassette.

Mark the tools which are used in the dialogue.

Compare your result with a partner.

3 Can I have a smaller spanner?

Listen to the cassette again.
What does Tina need the following tools for?
Connect the sentences.

She needs	a hacksaw	O	O	to take off the hoses.
	a small spanner	O	O	to cut off the hoses.
	a screwdriver	O	O	to get the radiator off.
	a drill	O	O	to put in the radiator.
	bolts	O	O	to make holes for the new radiator.

4 He needs a ...

Peter is changing the old exhaust on his car.
Which tool does he need for each step?
Complete the following statements.
Work in pairs.

To clean the pipes he needs a ....

To loosen the bolts he needs a ....

To cut the head off the bolt he ....

....

To get the bolt out of the holes ....

....

Describe a workshop project of your own.

Metals everywhere

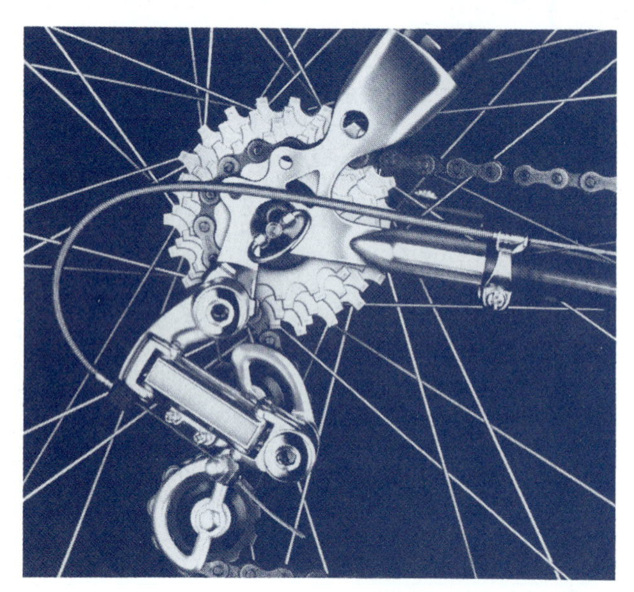

1 What's it called?

Look at the photos on this page and opposite and go through the following list of objects.
Which of the objects can you identify?

- ○ cylinder block ....
- ○ wheel rim ....
- ○ kitchen sink ....
- ○ cable ....
- ○ milling tool ....
- ○ trumpet ....
- ○ derailleur ....
- ○ wrought-iron gate ....

What is each of the objects made of?
What do you think?
Discuss your result with a partner.

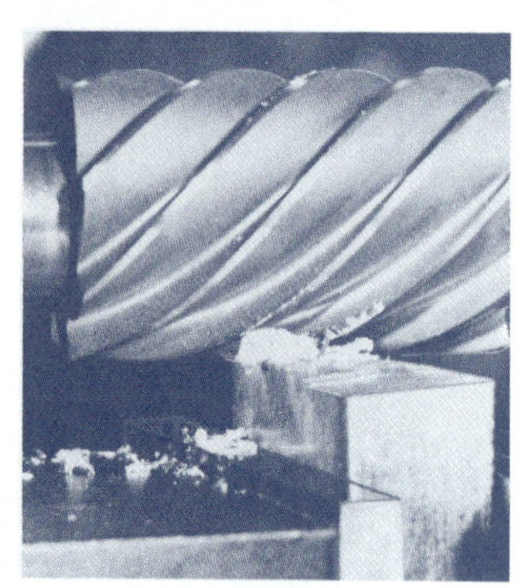

2 What's it made of?

Look at the photos and listen to the scene on cassette.
Which of the objects are the people talking about?

Look at the following list of materials and listen again.
What are the objects made of?

○ aluminium	○ copper
○ wrought iron	○ alloy steel
○ brass	○ plastic
○ stainless steel	○ cast iron alloy
○ titanium	○ high speed steel

Which of the materials is not a metal?
Compare your result with a partner.

3 Cast iron alloy?

Listen again and concentrate on the following part of the discussion.

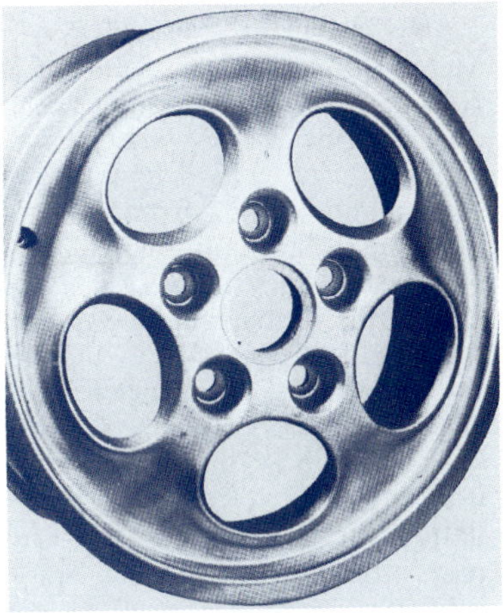

And this .... a cylinder block.

.... cylinders
Maybe a ....

What's ....

Cast ....

Could be.

Complete the text in the bubbles. Work with a partner.

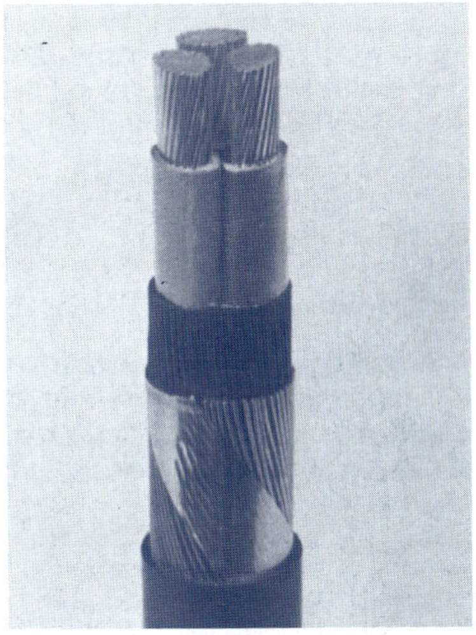

What are the things you work on in your workshop?

....

What are they made of?

....

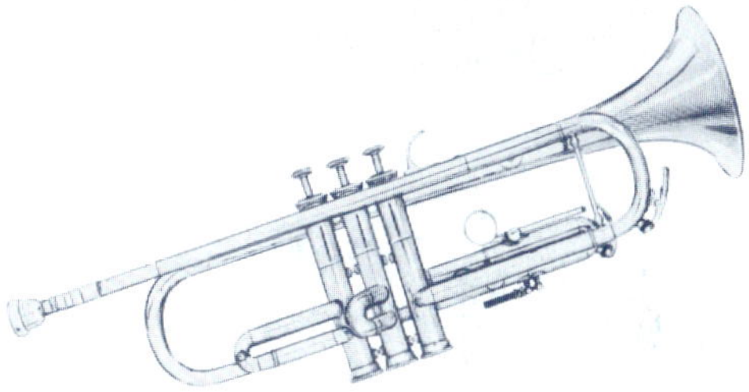

And what are your tools made of?

....

4 What's a car made of?

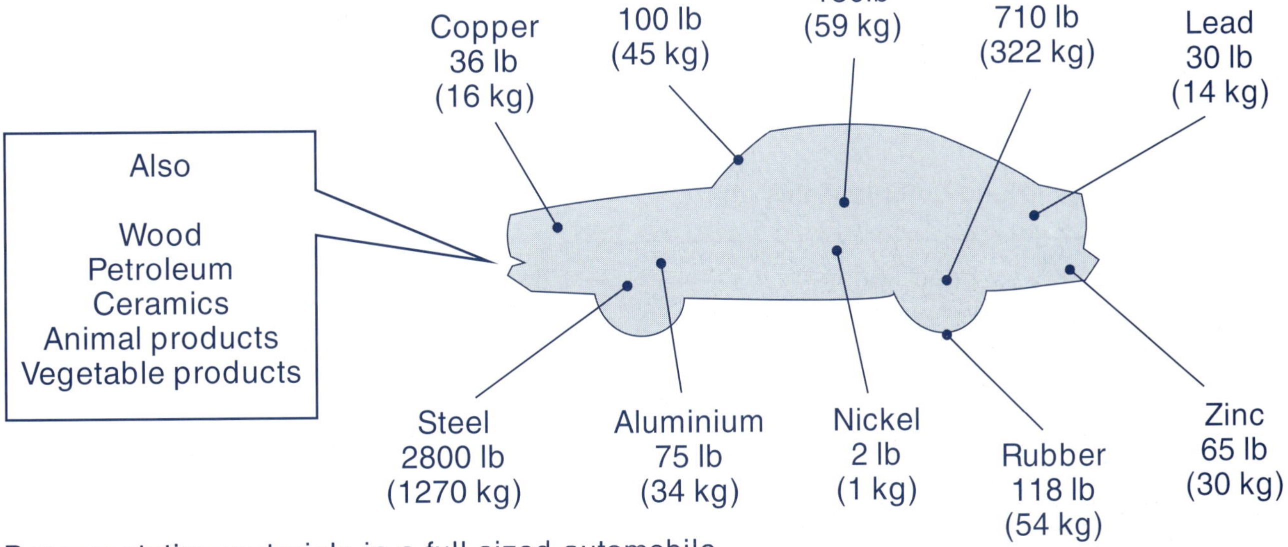

Also

Wood
Petroleum
Ceramics
Animal products
Vegetable products

Copper 36 lb (16 kg)
Glass 100 lb (45 kg)
Plastic 130lb (59 kg)
Iron 710 lb (322 kg)
Lead 30 lb (14 kg)
Steel 2800 lb (1270 kg)
Aluminium 75 lb (34 kg)
Nickel 2 lb (1 kg)
Rubber 118 lb (54 kg)
Zinc 65 lb (30 kg)

Representative materials in a full-sized automobile.

Metalworking bolts, washers and nuts. Hexagonal machine bolts, used in heavy-duty assembly, have hexagonal heads; common sizes include diameters from 6 to 12 mm and lengths from 25 to 150 mm. They are used to join two pieces of metal, when both sides are accessible. Set bolts are similar to machine bolts and come in the same sizes, but they are threaded along the entire body; they are used to achieve a close fit between two pieces of metal only one of which is threaded. Stud bolts are used to fasten two parts together in a situation where you may wish to remove one part without removing the bolt from the other part; they have an unthreaded centre and come in the same sizes as machine bolts. Coach bolts, which are used for joining metal to wood, have round heads with square collars at the top of the shank; the square collar bites into the wood to prevent the bolt from turning. They are available in the same sizes as machine bolts.
Common bolt nuts are hexagonal so they can be tightened with a spanner. A cap nut can be used to cover the free end of a bolt. Wing nuts are for parts that have to be disassembled frequently. The thin nut called a locknut is often used as a washer, to lock a bolt and nut in place; it may also be used alone for a tight but adjustable fit. Nuts are available in the same diameters as bolts. The two most common types of washer, the flat washer and the lock washer, are also available in standard sizes; lock washers are designed to press against the work and the nut.

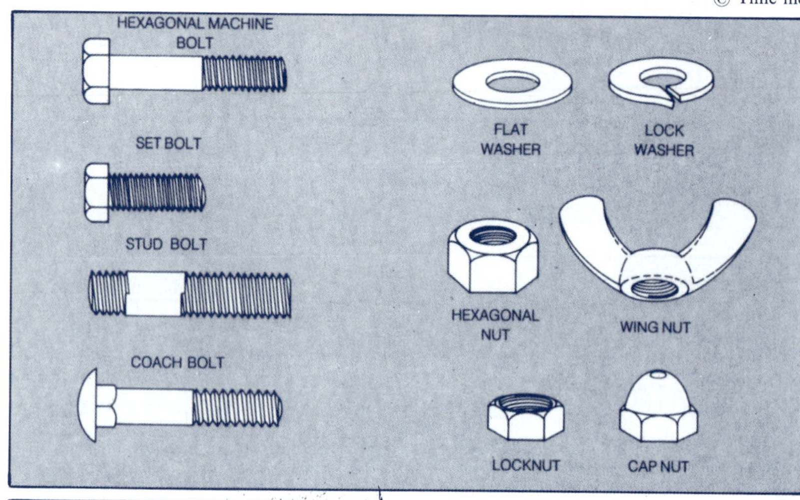

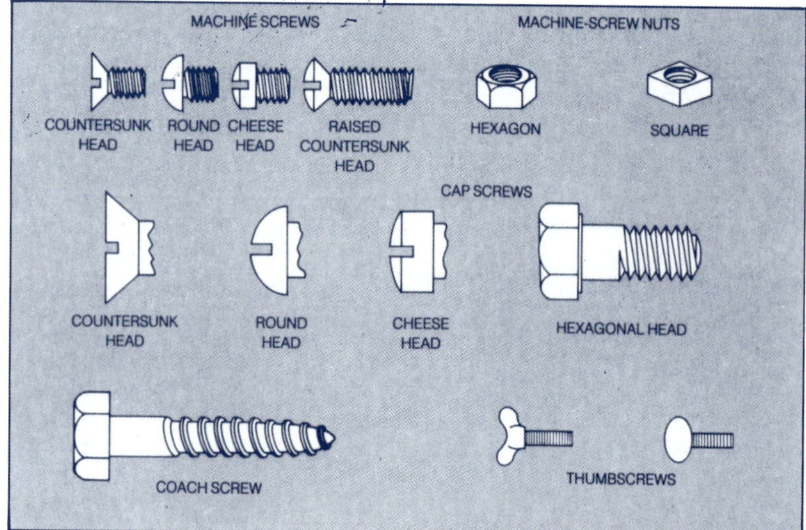

Metalworking screws. Machine screws have four head styles—round, cheese, raised countersunk and countersunk. The common sizes are No. 8 to No. 12; these range in diameter from 3 to 6 mm and from 12 to 50 mm long. Their hexagonal nuts are like those for bolts but smaller; square nuts are also used. Cap screws—threaded only part way—join two pieces where only one has a threaded hole. They range from 5 to 30 mm in diameter and from 12 to 150 mm in length. Thumbscrews, used when parts need to be disassembled, come in the same sizes as machine screws, with either a flat or winged head. Coach screws, used to attach metal to wood, have a bolttype head but are threaded like a screw. They come in the same sizes as coach bolts.

1 Screws and nuts

Read the text and underline all the screws and nuts you find.

2 Head styles

Read the text again and complete this chart.

	head style(s)	diameter	lengths
machine bolts			
cap screws			
machine screws			

3 What are they used for?

Read the text once more and connect the statements.

Screws or bolts are used . . .			
machine bolts	O	O	to achieve a close fit between two pieces of metal when only one of which is threaded.
set bolts	O	O	to join two pieces of metal when both sides are accessible.
stud bolts	O	O	to attach metal to wood.
coach bolts	O	O	to fasten two parts together in a situation where you may wish to remove one part without removing the bolt from the other part.
coach screws	O	O	to join metal to wood; they need a nut.
cap screws	O	O	when parts need to be disassembled.
thumbscrews	O	O	to join pieces when only one has a threaded hole.

What can you say about nuts and washers?
Go through the text again and connect the sentences.

common bolts	O	O	can be used to cover the free end of a bolt.
cap nuts	O	O	are hexagonal so they can be tightened with a spanner.
wing nuts	O	O	are often used as washers to lock bolts and nuts in place.
locknuts	O	O	are for parts that have to be disassembled frequently.
lock washers	O	O	are hexagonal like those for bolts but smaller.
machine-screws	O	O	are designed to press against the work and the nut.

4 Self-tapping or self-drilling?

Look at the drawing and listen to the instructions.
Which types of screws are mentioned?
Tick the right one(s).

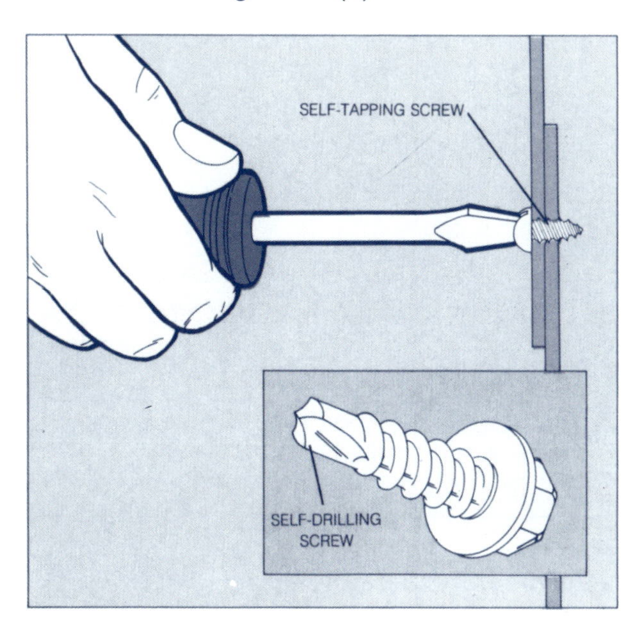

Listen again and tick the right statements

A self-tapping screw
- O drills its own hole.
- O cuts its own thread.
- O doesn't have a thread.

The diameter of the hole should be
- O bigger than
- O the same as
- O smaller than

} the screw shaft.

A self-drilling screw can be driven in
- O with a power drill.
- O with a screwdriver.
- O with a nut-driver.

Think of all the steps it takes to join two metal parts with nuts and bolts or screws.
Then describe the process step by step.

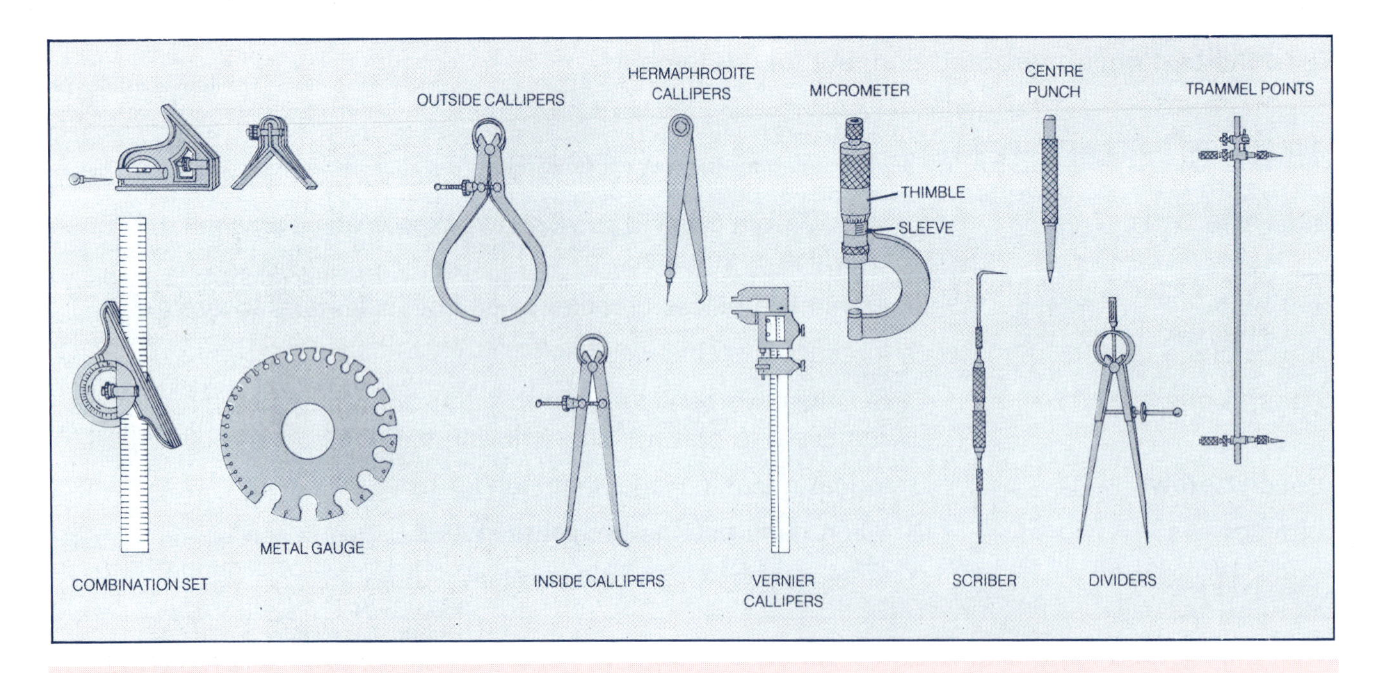

In addition to the standard measuring tools—such as steel rules and squares—there are a number of other useful measuring and marking tools.
One is the combination set. It has three interchangeable heads that can be mounted on a single steel rule. One of these heads is used for marking off angles from 0 to 180 degrees; the second one is used for finding the centre of a cylindrical shaft; and the third one is for checking metal corners for squareness. Also useful is the metal gauge. By inserting metal stock into the slot that fits it best, you can identify the gauge of that piece of metal.
To measure very thick stock or irregularly shaped objects you can take various types of callipers—inside and outside callipers. They can be used with a steel rule for simple measurements of width or diameter.
Vernier callipers combine inside and outside measuring and contain a built-in rule. Measuring thickness or diameters of metal pieces to the nearest hundredth and sometimes even two-thousandth of a millimetre you can use an outside micrometer.
A good tool to use for laying out measurements or designs on metal is a scriber. To mark a single point, a hardened steel centre punch is useful. And for laying out small circles and arcs, use steel-tipped dividers. For scribing larger circles, a pair of trammel points is the most convenient tool.

1 Measuring

Read the text and underline the tools for measuring and marking.
Go through the text again and answer the following questions.

What do you need . . .	Tools
. . . to mark a single point on metal?	....
. . . to scribe a large circle?	....
. . . to scratch a fine line on metal?	....
. . . to measure fractions of a millimetre?	....
. . . to measure an inside width?	....
. . . to measure very thick stock?	....
. . . to identify the gauge of metal stock?	....

2 Making a notch

Listen to the instructions on cassette and look at the pictures. Can you number the pictures in sequence?

Listen again and number the instructions in the right sequence.

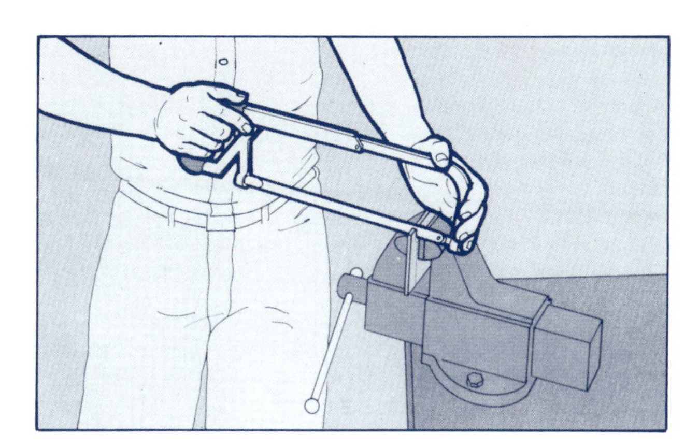

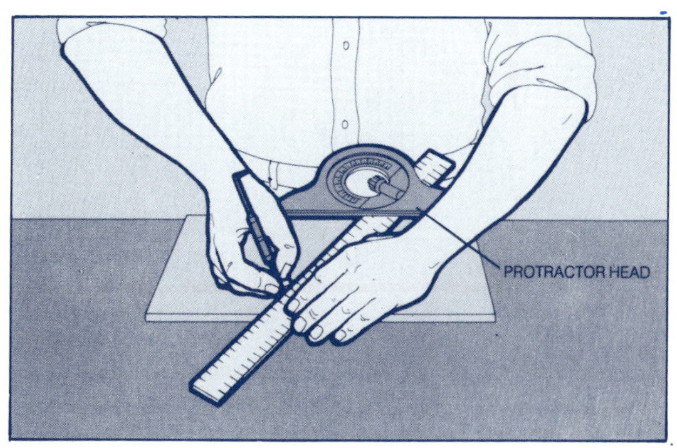

○ Mark the lines.

○ Scribe the cutting lines.

○ Measure the lines for cutting.

○ Smooth with a file.

○ Clamp the metal stock in a vice.

○ Cut the base of the notch with a chisel.

○ Cut the sides of the notch with a hacksaw.

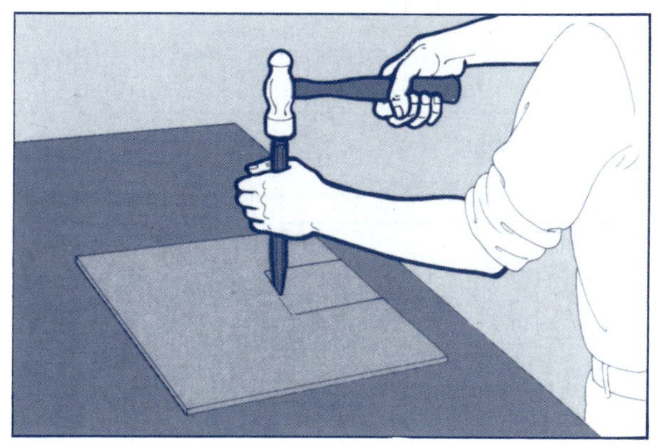

3 All the tools

Listen again:
What do you need to make a notch?
Connect the boxes.

	○ a micrometer
For measuring ○	○ a scriber
	○ a vernier calliper
	○ a combination set
For marking ○	○ trammel points
	○ a hacksaw
	○ a chisel
	○ dividers
For cutting ○	○ a centre punch
	○ a circular saw
	○ a file
	○ tin-snips
For smoothing ○	○ sand-paper
	○ a grinding wheel

Which measuring tools are used in your workshop or company?
What are they used for?

Cutting chips

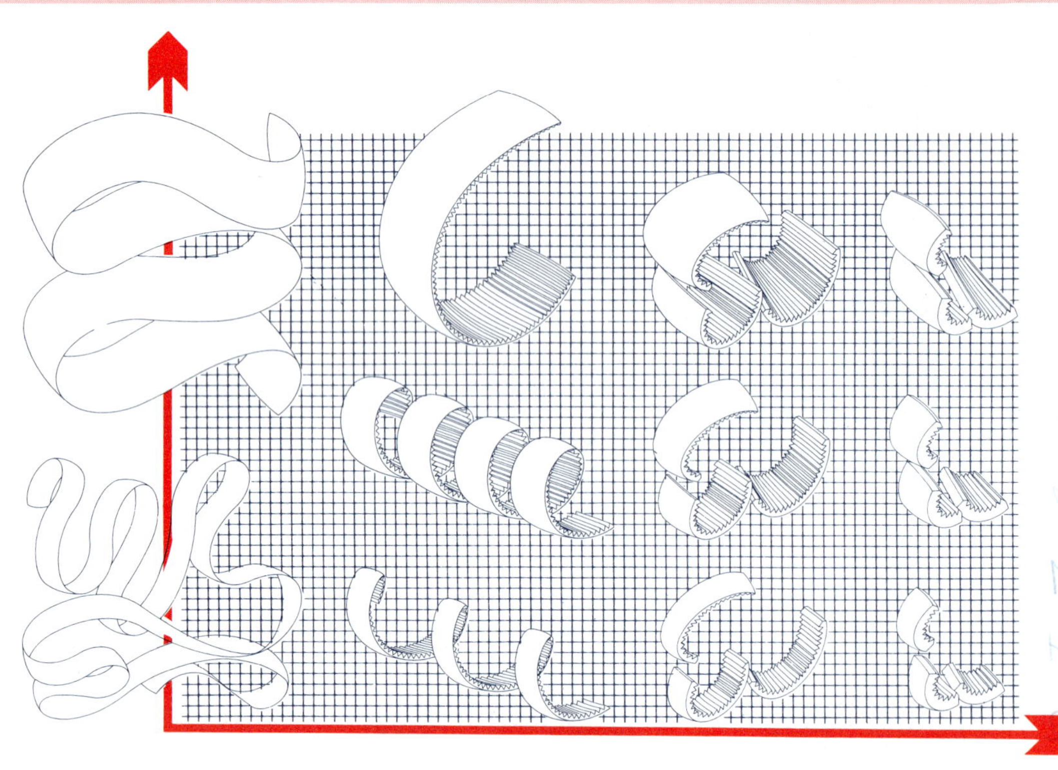

Fish and chips,
casino chips,
electronic chips,
metal chips,
potato chips,
cutting chips
or
computer chips?

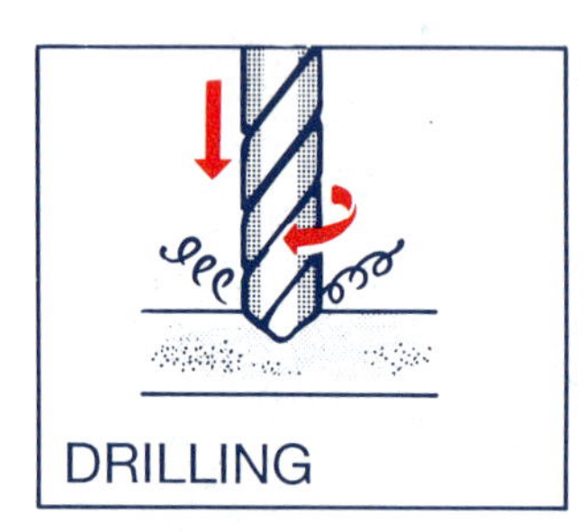

....

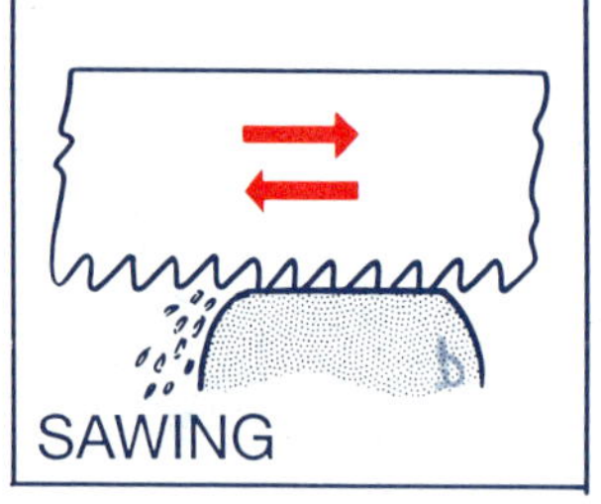

....

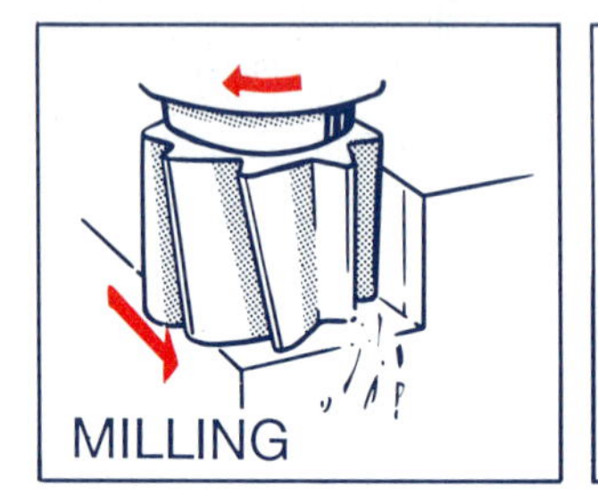

....

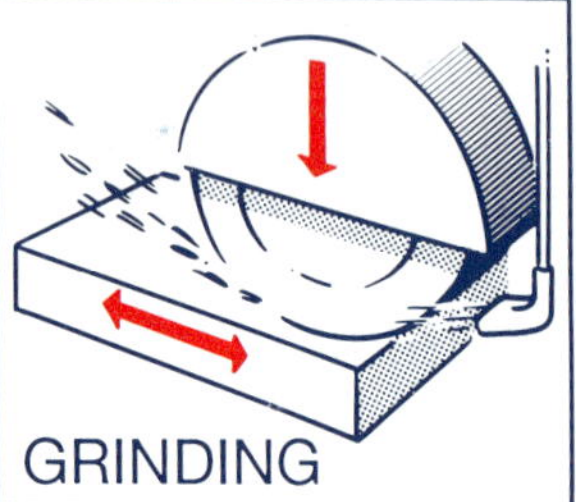

....

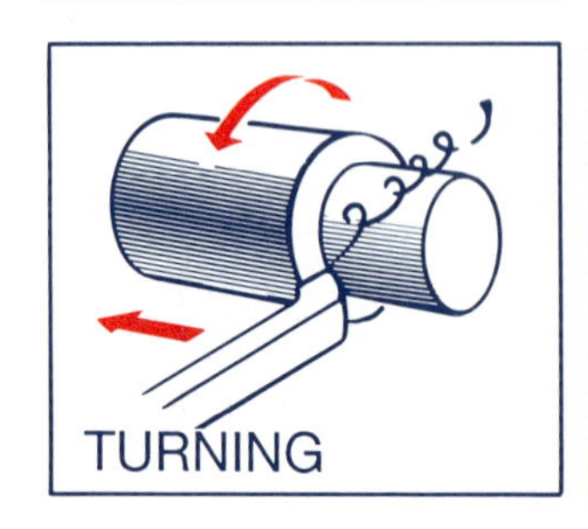

....

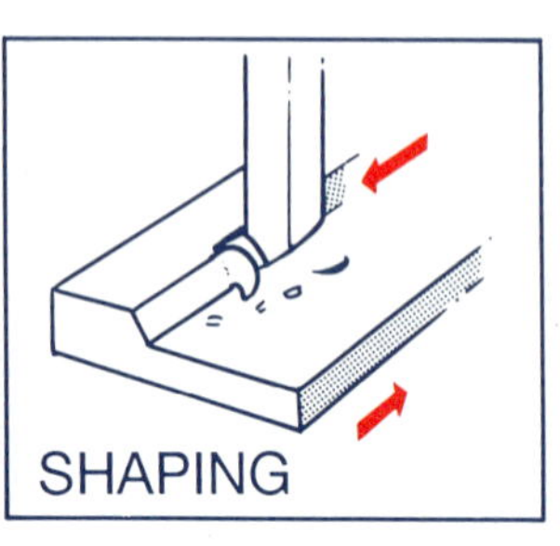

....

1 What kind of chips?

What kind of chips do you get from hand-held filing, from sawing, drilling, shaping, turning, milling, and even from grinding?

2 Chipping action

Look at the drawings on the left and at the machines on the opposite page. What is being done on which of the machines? Write the name of the machine below each picture of a process.

3 A machine tool operator

Look at the drawings and listen to the interview. What does a machine tool operator do? Tick the machines the man works on.

Listen again and complete the bubbles.

What does .... do?

I can work on all the machines you see in here, like on the .... over there, on the .... in the back, and on the ...., of course.

1

This is a process used mainly for producing round bars and cylinders. The workpiece is rotated against a cutting tool, metal being removed by peeling action.

2

This is a process principally used for producing flat surfaces but may be used to make special shapes. The workpiece is moved slowly under a rotating tool which has several cutting edges. Metal is removed by chipping action.

3

This is a process used for producing smooth flat surfaces. Metal is removed in small quantities by passing the workpiece across a rapidly rotating abrasive wheel. The process is often used to bring items to their finished size after hardening.

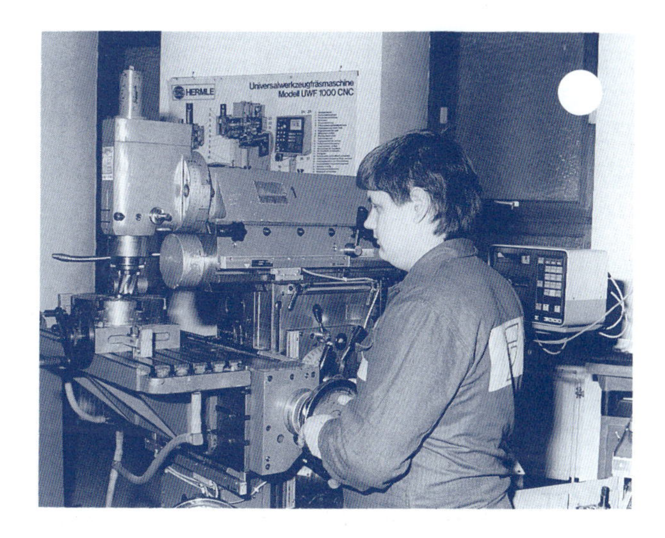

4 Machines and processes

Look at the photos and read the texts.
Which machine is used for which process?
Number the photos,
then write a headline for each text.

5 Which machines can you use?

Which machine would you use . . .

. . . to drill holes? ....

. . . to produce round bars? ....

. . . to produce flat surfaces? ....

. . . to make special shapes? ....

. . . to produce smooth surfaces? ....

I'd use a surface grinder.

Which machine(s) have you worked on?

I've worked on a Xerox copier.

1 Drilling machines

Look at the photo and at the drawings on this page and on the next.
Which kind of drilling machines are presented?

- ○ bench model
- ○ pillar drill (press)
- ○ radial arm drill
- ○ floor model
- ○ electric drill
- ○ upright drill press

MAKITA - INDUSTRY MODELS WITH PISTOL GRIP
with belt suspension, fully fitted with ball bearings, complete with drill chuck
6510 B - 2: with mechanical 2 - step gear - box
6510 LVR/DP 4700: with special low - speed power transmission and infinitely variable electronic, righthand - lefthand switchable

Art. No.	Model	Drilling capacity Steel	Wood	Power input/W	R.p.m.	Price
4051 017	6500 PB	6,5	13	280	2300	
4052 013	6510 PB	10	15	330	1800	
4073 011	6510 B - 2	10	20	330	950/2300	
4056 019	6510 LVR Electronic R+L	10	15	400	0 - 1050 right.+lefth.	
4087 011	DP 4700 Electronic R+L	13	30	510	0 - 550 right.+lefth.	

MAKITA - INDUSTRY MODEL WITH BACK GRIP
fully fitted with ball bearings, complete with drill chuck
6300/4 (4 speed)

Art. No.	Model	Drilling capacity Steel	Wood	Power input/W	Rp.p.m.	Price
4045 017	6300/4 (4 speed)	13	30	650	500/600 1100/1300	

2 Electric drills

Look at the tables from the tool catalogue.
Which of the drills listed is the one in the photo above?

Test your neighbour, like this:

They've got an electric drill with variable speed from zero to one thousand and fifty rpm.
It can drill 10mm of steel or 15mm of wood.
Which one is it?

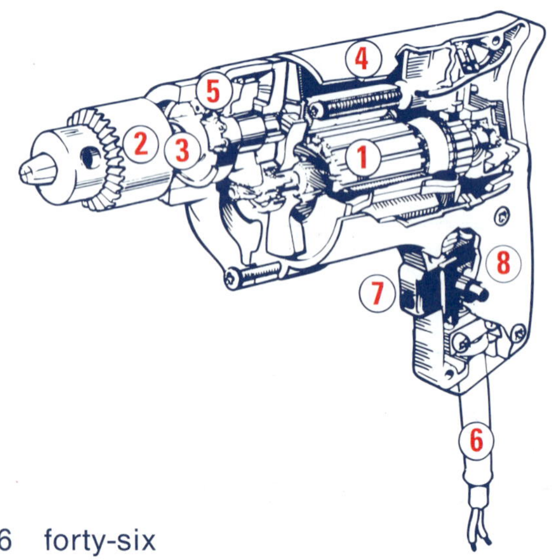

Look at the drawing and number the parts list.

power cord	○	chuck	○
ball bearing	○	spanner flats	○
motor	○	trigger switch	○
handle	○	motor frame	○

3 A pillar drill?

Look at the drawings on page 56 and 57 and listen to Tina.
She's showing some visitors around the machine shop.
Which of the drilling machines does she mention?

....

Which machine does Peter work on?

....

Listen again and look at the parts list.
Which parts does Peter mention?

- ○ pillar
- ○ column
- ○ drilling head
- ○ spindle
- ○ base
- ○ clamping levers
- ○ table
- ○ V-belt
- ○ motor

Listen again.
This time concentrate on how Peter describes his drilling machine.
Complete the bubbles.

A pillar drill? Sounds Greek to me...

Well, most upright drill presses have a .... or ....

Look here: this is the ....
It's mounted ....,
and it holds ....
for the workpieces to be drilled.

And on top of the ....
that's the ....

Compare your result with a partner.
Can you now label the drawing of the pillar drill?
Explain it to your partner.

4 A radial arm drill

Look at the drawing of the radial arm drilling machine.
Can you complete the missing words?
Explain the machine to a partner.

What kind of drilling machine(s) have you got at your company?
And at the workshop in your school or college?

Describe the machine to a partner or tell your class.

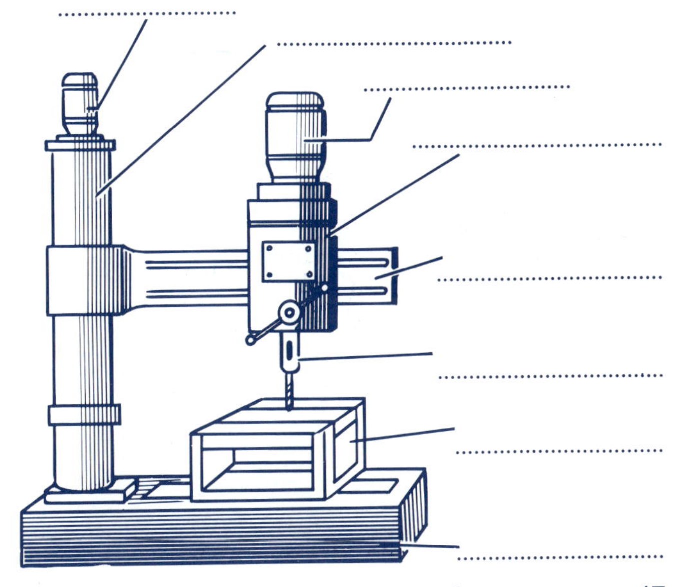

The major parts of the lathe are the bed, the headstock, the carriage and the tail-stock.
The bed is the back-bone of the lathe. It carries all the other main parts.
The headstock is the brain of the lathe. It contains the main gear driving the spindle with the chuck, the feedgear driving the lead screw or the feed rod, and the speed and feed control panels. The carriage is the centre of all lathing operations. It includes the saddle and the apron, and carries the cross slide and the tool rest. In the tailstock a centre can be mounted for working longer workpieces. Both tailstock and carriage ride on the bed of the lathe.

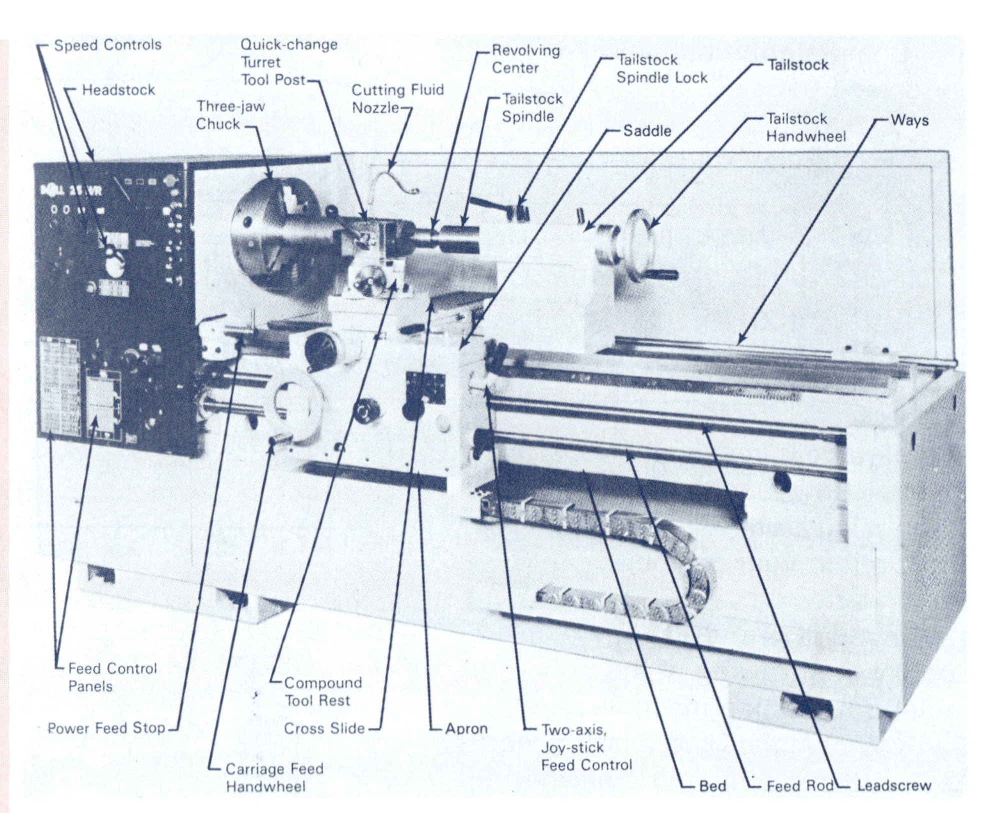

1 Manual or numerical control?

Look at the photos above. They show a manually operated lathe and a numerically controlled one in operation. Which is which?

2 Bed, carriage, head- and tailstock

Look at the photo and study the parts of the lathe.
Read the text and underline the main parts that are described.

Boring
is the machining of a hole already drilled. Most of the above mentioned external operations are performed internally as well with tools specially developed for the purpose.

Parting off (Cutting off)
Instead of sawing off the component a special parting tool is used to cut off the component from the workpiece bar in the lathe. It uses a thin bladed tool which cuts towards the centre of the workpiece.

Threading
This can be performed with a tool having the same form as the thread. The tool is then fed along at a rate which corresponds to the pitch of the thread. (Screw threads can, however, also be produced using taps and dies.)

Facing
is also a common operation where the tool turns a face perpendicular to the workpiece axis, either away from centre or in towards centre.

Longitudinal turning
is the most common turning operation where the tool moves along the workpiece axis reducing the diameter.

Form cuts
are performed with tools which have been shaped to the specific form. The most common are various types of grooves, recesses or chamfers.

Out-copying and in-copying
can be performed at various angles and even along various radii. Some workpieces consist of combinations of these cuts and acute angles which make demands on the accessibility that can be obtained with the tool.

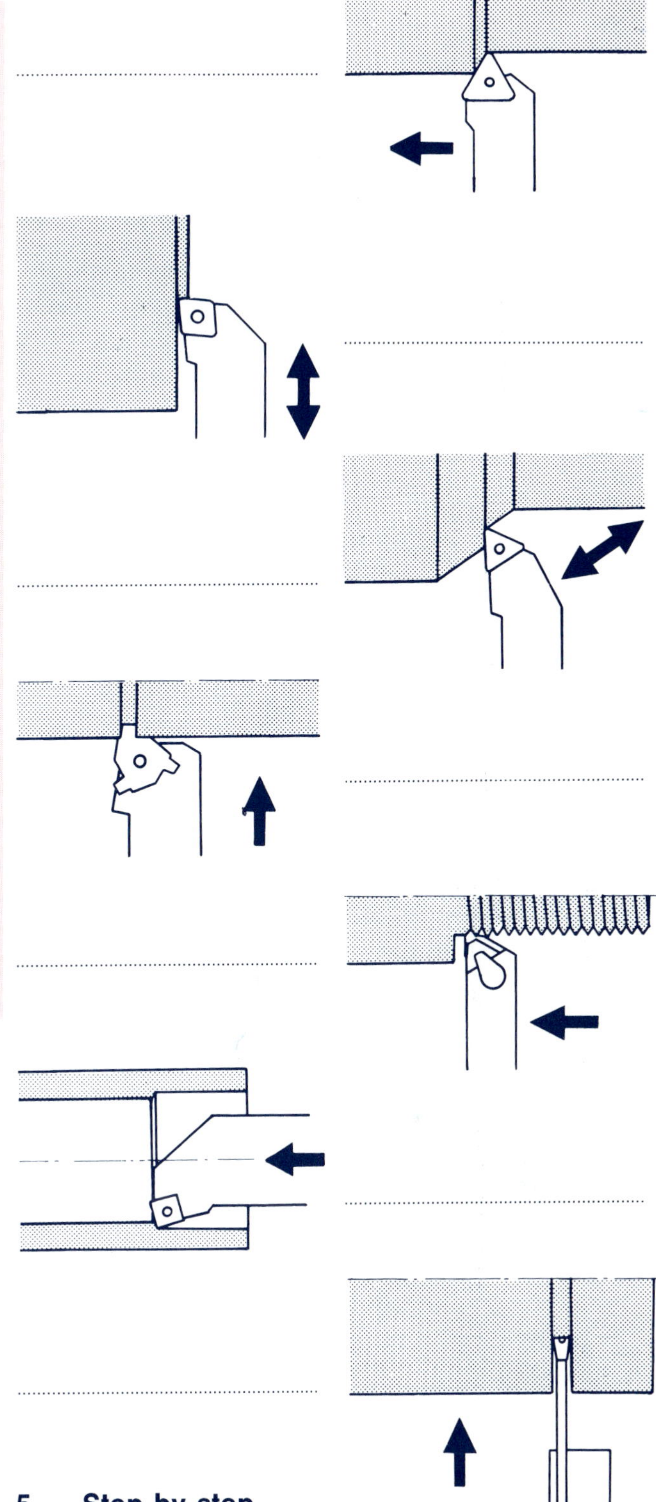

3 Basic operations

Look at the drawings and read the descriptions of the various types of operations.
Which description belongs to which picture?
Label each drawing with the right operation.

4 Reduce the diameter

Look at the drawing and listen to the dialogue.
Can you fill in the missing diameters?

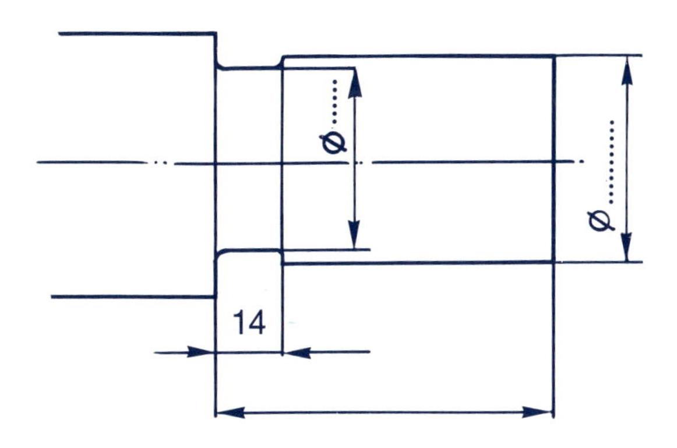

5 Step by step

Listen again and number the steps.

◯ Reduce the diameter.

◯ Face the bar.

◯ Cut the thread.

◯ Cut the groove.

Describe the lathing operation in your own words.

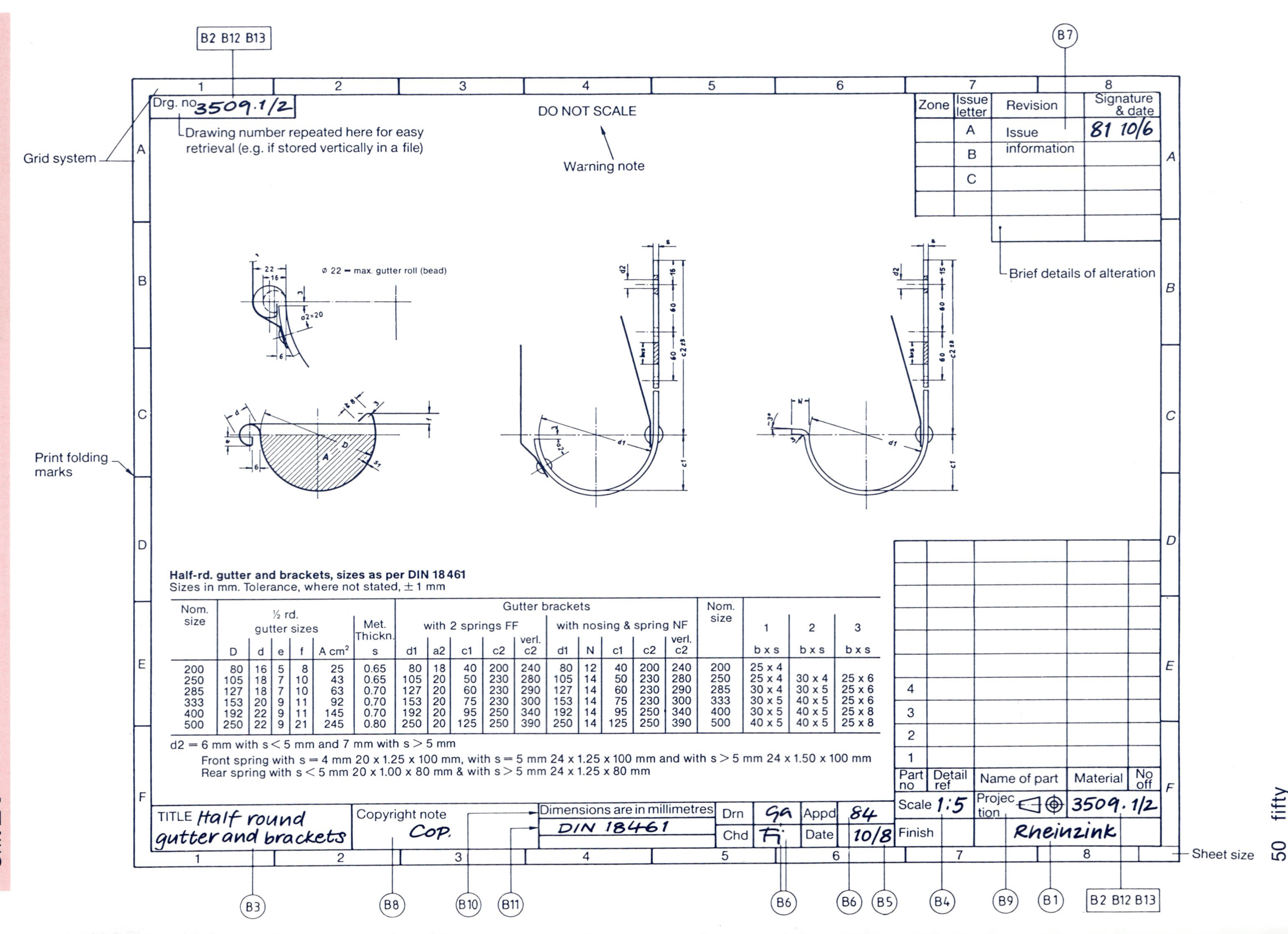

Half-rd. gutter and brackets, sizes as per DIN 18 461
Sizes in mm. Tolerance, where not stated, ± 1 mm

Nom. size	½ rd. gutter sizes					Met. Thickn.	Gutter brackets with 2 springs FF					Gutter brackets with nosing & spring NF					Nom. size	1	2	3
	D	d	e	f	A cm^2	s	d1	a2	c1	c2	verl. c2	d1	N	c1	c2	verl. c2		b x s	b x s	b x s
200	80	16	5	8	25	0.65	80	18	40	200	240	80	12	40	200	240	200	25 x 4		
250	105	18	7	10	43	0.65	105	20	50	230	280	105	14	50	230	280	250	25 x 4	30 x 4	25 x 6
285	127	18	7	10	63	0.70	127	20	60	230	290	127	14	60	230	290	285	30 x 4	30 x 5	25 x 6
333	153	20	9	11	92	0.70	153	20	75	230	300	153	14	75	230	300	333	30 x 5	40 x 5	25 x 6
400	192	22	9	11	145	0.70	192	20	95	250	340	192	14	95	250	340	400	30 x 5	40 x 5	25 x 8
500	250	22	9	21	245	0.80	250	20	125	250	390	250	14	125	250	390	500	40 x 5	40 x 5	25 x 8

d2 = 6 mm with s < 5 mm and 7 mm with s > 5 mm
Front spring with s = 4 mm 20 x 1.25 x 100 mm, with s = 5 mm 24 x 1.25 x 100 mm and with s > 5 mm 24 x 1.50 x 100 mm
Rear spring with s < 5 mm 20 x 1.00 x 80 mm & with s > 5 mm 24 x 1.25 x 80 mm

1 Basic information

Study the drawing on the opposite page and complete the following chart.
Work with a partner.

Ref. No.	Basic information	What do you need to know?	Actual information
B 1		Who is the drawing from?	
B 2	Drawing number		
B 3		What does it show?	
B 4	Original scale		
B 5		When was it drawn?	
B 6	Signatures	Who drew it? Who checked/approved it?	
B 7		Is this the latest issue?	
B 8		Is it confidential? Or may it be copied?	
B 9		What type of projection has been used? First/third angle?	
B 10		Which measurement units have been used? Metric/imperial?	
B 11	Standards	Which standards have been used?	
B 12	Sheet number	Which sheet is it?	
B 13	No. of sheets	How many sheets are there?	

2 We've got the gutters

Look at the drawing on the opposite page and listen to Susan and Peter.

Which parts are they talking about?

....

Which part do they need?

....

3 Dimensions

Look at the drawing and study the table of dimensions.
Then listen to Susan and Peter again.
What size do they need?
Mark the dimensions they discuss.

4 Symbols

Listen once more:
What do the following symbols stand for? Connect the boxes.

D ○	○ gutter roll diameter
d ○	○ gutter diameter
d1 ○	○ length of bracket arm
c2 ○	○ bracket inside diameter
s ○	○ fastening hole diameter
b ○	○ thickness of bracket
d2 ○	○ width of bracket

Compare your result with a partner.

CADLINES

1 The components

What does CAD stand for?
What do you think?

- ○ Control and despair.
- ○ Computer aided design.
- ○ Cash and dine.

Look at the computer components you need for CAD.
Match the names and the photos.

1. computer
2. keyboard
3. graphic tablet
4. mouse
5. monitor
6. plotter

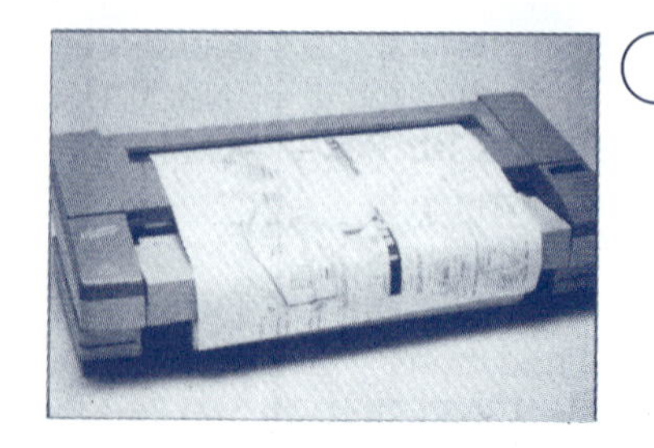

2 The functions

What do you know about CAD?

Look at the computer components and listen to the discussion.

Tick the components they mention.

Listen again and connect the components and their functions.

The computer ○	○ shows what you draw.
The monitor ○	○ does all the calculations.
The keyboard ○	○ is used to feed the necessary information.
The graphic tablet ○	○ translates drawings for the computer.
The plotter ○	○ transfers your work on paper.
The scanner ○	

Compare your results with a partner.

3 What about you?

Have you already worked with CAD?
Have you seen CAD in action?
What did you use it for?
What did you draw on it?

THE ONLY RASTER PLOTTER THIS FAST WITH A PRICE TAG THIS SMALL.

Increased Productivity
Thermal plotting plus our built-in FastPlot protocol give you faster file transfers and lightning-quick output. Output time: 1 minute.

Variety of Media
ExpressWrite™ extra-long D-size roll media (premium, report grade and film) yields more plots for less money than other thermal plotters.

Easy to Use
Configure the plotter with the easy-to-read control panel, or the screen-based ExpressPanel™ — right from your PC or Sun® workstation.

Dependable Technology
Thermal technology is quiet, trouble-free and environmentally safe — no pens, toners or chemicals.

Better Image Control
Plot tiling (nesting), auto rotation, eight line widths, 50% line screen, and more.

Hands-off Plotting
With convenient roll feed and the ExpressPanel™, individual or networked users can plot continuously and unattended — no more "babysitting" the plotter.

Call toll-free for more information about the very affordable, incredibly fast ExpressPlotter™ and the JDL family of plotters. And ask for your free Plotter Selection Guide. Or FAX the coupon to us today.

FAX (805)388-8708

For the name of your nearest dealer, call

(800)624-8999, ext. 246

In Canada, call (416)675-3999

Express Plotter

JDL
Japan Digital Laboratory

☑ *Please send my Plotter Selection Guide right away!*

Name ____________

Company ____________

Address ____________

City ____________ State ______

Zip ________ Phone (_____) ____________

Mail to: JDL, 4770 Calle Quetzal, Camarillo, CA 93012
Call: 805-388-8709 or Fax 805-388-8708

FEDERAL AGENCY OR DOD AFFILIATED READER SERVICE NO. 101 • DEALER INQUIRIES READER SERVICE NO. 102 • ALL OTHER INQUIRIES READER SERVICE NO. 103

4 Quiet, trouble-free and environmentally safe

Read the advertisement. Concentrate on how the features (nouns) are described. Write the adjectives in front of the nouns.

adjective	noun	adjective	noun
....	productivity	....	control panel
....	plotting	....	roll feed
....	file transfers	....	users
....	output	....	image control

And .... plots for .... money.

Which characteristics would you need in a plotter? Mark them and discuss them with a partner.

A Kangofant?

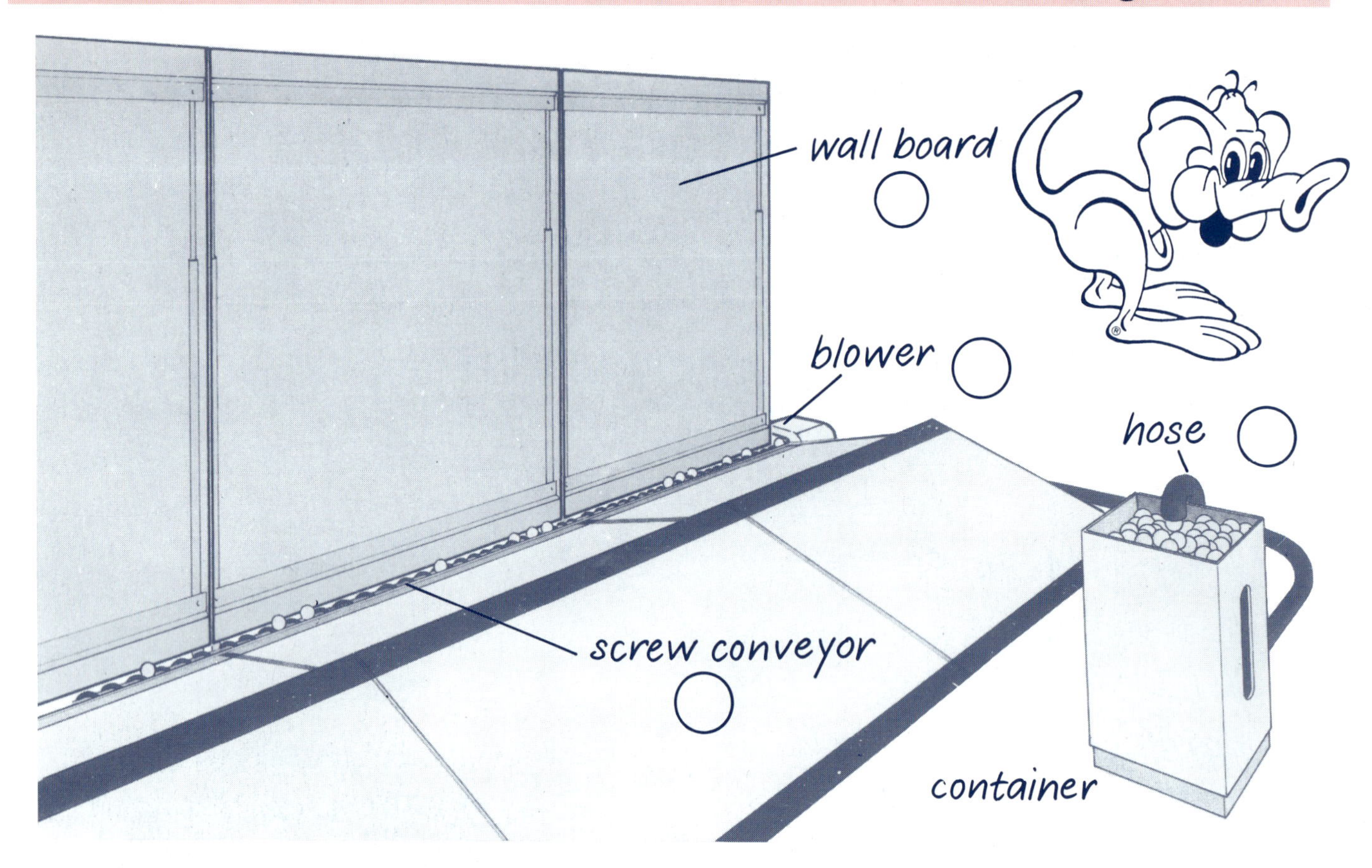

1 Like kangoroo and elephant?

Look at the picture and listen to the dialogue.
What is a Kangofant?

- ○ A cross between a kangoroo and an elephant.
- ○ A tennis ball collecting machine.
- ○ A tennis ball throwing machine.

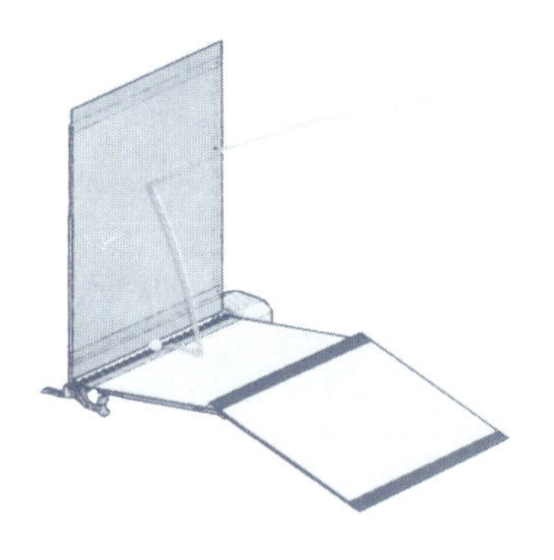

Now listen for the elements of the Kangofant.
Number them in the sequence they are mentioned.

Compare your result with a partner.

2 Blown through a hose

Listen to the dialogue once more
and complete the function of the Kangofant.
Use the following verbs:

blow (blown), pick up, deflect, play, feed (fed).

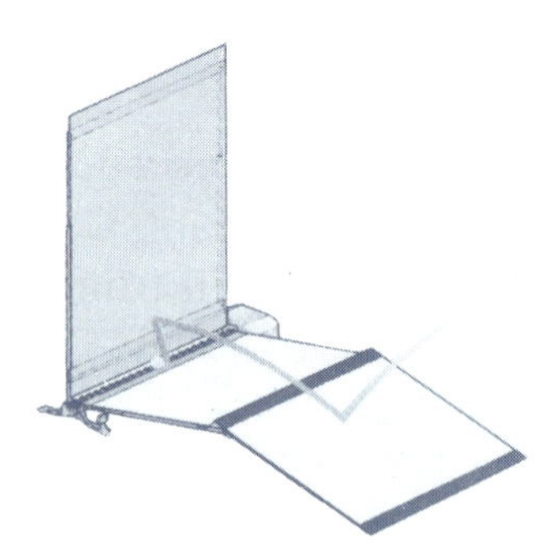

You .... the balls.
From the wall boards they are .... on to a screw conveyor.
They are then .... to a blower,
and .... through a hose to the container.
There you .... your new balls.

3 A tennis ball collecting machine

Read the text and find out the technical data.

Collecting capacity ....

Power consumption ....

Wall board sizes ....

The Kangofant is a tennis ball collecting machine with a collecting capacity of 1600 balls per hour. It consists of deflecting walls, a screw conveyer, a central driving unit with a blower, a hose and a tennis ball container/dispenser.
The wall boards are installed at one end of the tennis court and deflect the balls towards the screw conveyer. The screw feeds the balls into the blower which propels them through the hose into the container or into a ball throwing machine.
The screw conveyer and the blower are driven by small three-phase electric motors, with a total power consumption of 170 Watts. Up to nine wall boards, each 2 metres wide and 2.60 metres high, can be combined to cover the width of a tennis court.
The individual elements are colour-coded and easy to assemble, no tools required.

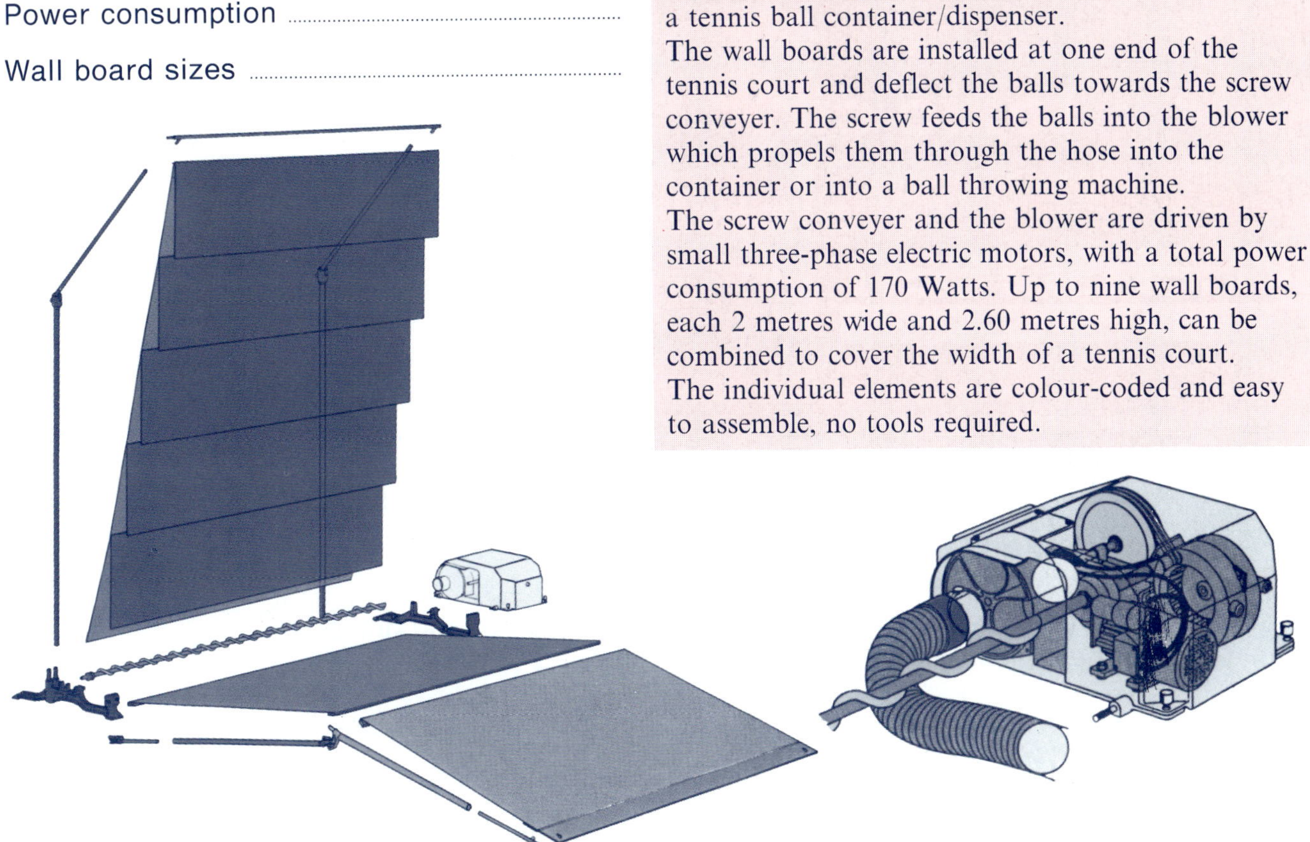

Read the text again and connect the following sentences.

The wall boards	○	○ feeds the balls into the blower.
The screw conveyer	○	○ deflect the tennis balls.
The blower	○	○ feeds the balls into a container.
The hose	○	○ propels the balls through a hose.

Explain the function and data to a partner.

4 A power station

Look at the picture and complete the description of the power station. Use the verbs below in the passive form. Work with a partner.

convert, produce, pump, use.

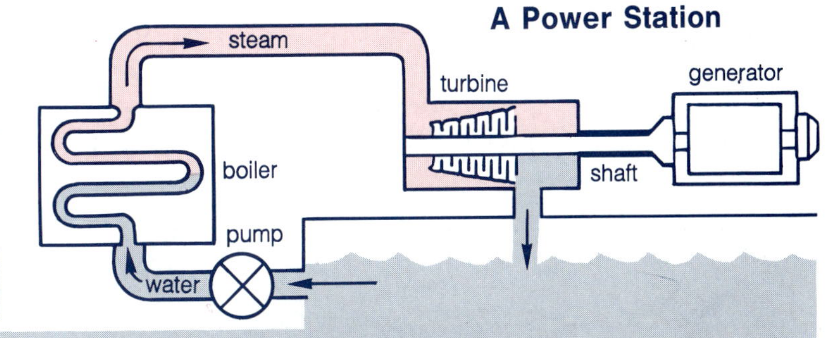

Water is *pumped* along the boiler pipes. In the boiler water is .... into high-pressure steam. In the turbine steam is .... to drive the generator shaft. With the generator electricity is ....

Describe the function of another (simple) machine.

1 A lot of waste

Listen to the interview.

How do we create waste?

- ○ We buy products.
- ○ We use products.
- ○ We throw them away.

What do we do with the waste?

- ○ Bury it in landfills.
- ○ Burn it in incinerators.
- ○ Dump it at sea.
- ○ Export it.

What should we do instead?

- ○ Import waste.
- ○ Reduce buying products.
- ○ Reuse products.
- ○ Recycle waste.

Compare your results with a partner.

2 Waste kills

Listen to the interview again.
Find out about the problems with landfills, incinerators and dumping.
Connect the sentences.

Dumping at sea ○	○ kills seabirds and marine mammals.
Incinerators ○	○ produce smell and toxic chemicals.
Landfills ○	○ produce ash containing heavy metals.

© ANT, Australia

Listen again and fill in the gaps.
Use the following words:

> marine mammals, plastic debris,
> seabirds, whales, eat, entangle.

.... dumping into the sea is no solution either.

Of course not!
Every year more than ....
and over ....
and other .... die

They die because they become
....
or because ....

3 New environmental awareness

What can we do to develop an environmentally conscious lifestyle?
Read the following advice. Tick what you already do.

	At home	At work
Buy fresh food that doesn't need a lot of packaging.	○	○
Use bottles more than once, or take them to a bottle bank.	○	○
Save paper; buy and use recycled paper as often as possible.	○	○
Avoid "throw-away" products.	○	○
Try to cut rubbish by 50%.	○	○
Reuse packaging such as plastic bags.	○	○
Collect and recycle metals and chips.	○	○
Reduce and recycle coolants and lubricants.	○	○
Return rubber tyres to the manufacturer.	○	○
Dispose of paints properly.	○	○
Use non-toxic substances.	○	○
Use batteries as little as possible. It takes 50 times more energy to make them than they produce.	○	○

4 What we do

Discuss your ideas about reduce—reuse—recycle in groups of three or four.

I buy fresh food .....

We return most bottles to the shop. The rest we take to the bottle bank for recycling.

At my company we separate metals.

But we should be more careful with motor oil.

....

© Debbie Beevor

5 What your local authorities do

What happens to the rubbish that is collected in your area?

How much of it is dumped, incinerated or recycled?

Up to 4000 °C

1 Welding is dangerous

Listen to John talking about his job.

Which types of welding are used in his company?

- ○ oxyacetylene gas welding
- ○ ultrasonic welding
- ○ arc welding
- ○ electro-beam welding
- ○ spot welding
- ○ pressure welding
- ○ electro-slag welding
- ○ MIG-welding

2 Protect yourself

Look at the photo:
How does John protect himself?

He wears
- ○ a mask.
- ○ a hard-hat.
- ○ heatproof gloves.
- ○ a leather apron.
- ○ a face shield.
- ○ goggles.
- ○ safety shoes.

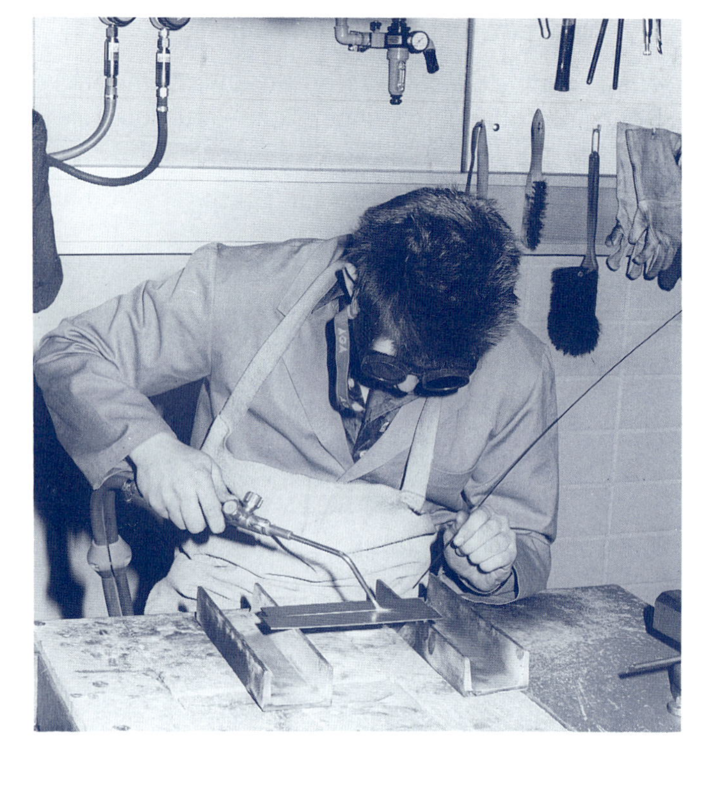

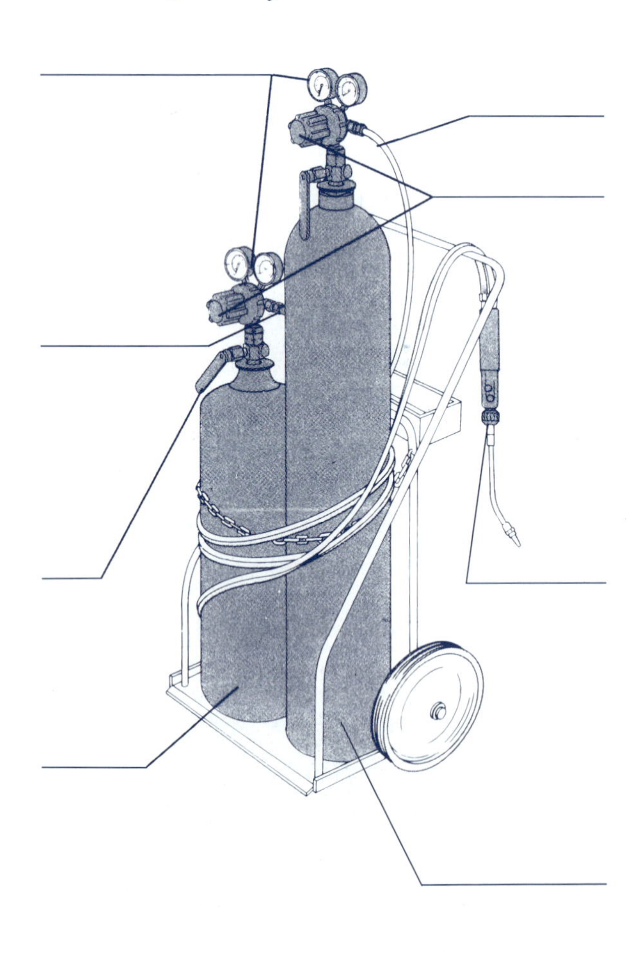

3 An oxyacetylene set

Listen to the discussion and tick the expressions you hear.

- ○ tip
- ○ tall cylinder for oxygen
- ○ spindle keys
- ○ torch with valves
- ○ two-wheeled handcart
- ○ short cylinder for acetylene
- ○ adjusting knobs
- ○ dual regulators
- ○ hoses: red for ....

  blue/black ....

Can you find out what each part is called in German?

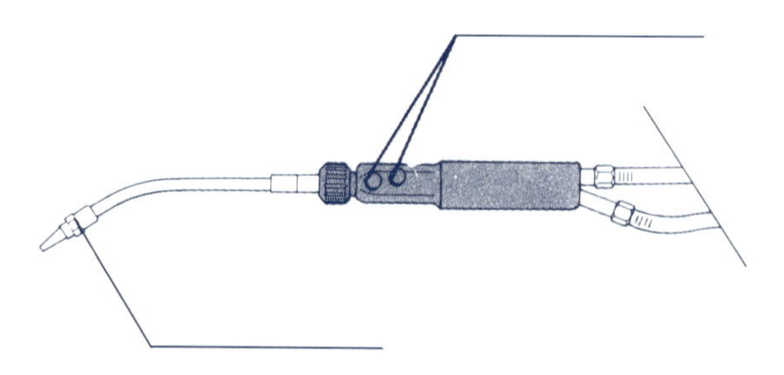

4 Preparing the set

Look at the pictures on the opposite page and read the texts.
Which picture belongs to which text?
Write the letters into the circles.
Then explain to a partner how to prepare the oxyacetylene set.

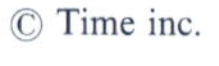

A Clearing the valves
For new cylinders, remove the protective caps and stand well clear of the valve outlets. Open the valve slightly by turning the spindle key a quarter turn anti-clockwise.

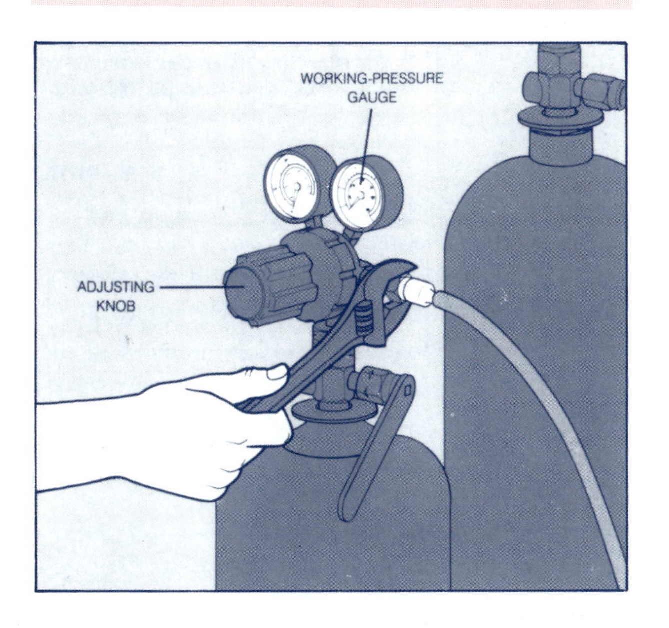

D Assembling the torch and tip
Connect the red hose to the torch fitting with the left-hand thread and the blue or black oxygen hose to the fitting with the right-hand thread. Tighten both of the nuts with a spanner.

E Checking for leaks
Open both cylinder valves. The oxygen working-pressure gauge should read 20 pounds (1.3 bar) and the acetylene one 5 pounds (0.3 bar). Close both cylinder valves and watch the cylinder-pressure gauges; if they fall, there is a leak.

B Attaching the regulators
Make sure the adjusting knob is closed fully. Put the regulator fitting inside the valve nozzle and tighten the regulator nut anticlockwise, first by hand, then with a spanner, until it is tight.

C Attaching the hoses
Connect the red hose to the acetylene regulator and the blue or black one to the oxygen regulator. The acetylene hose is fitted onto the hose connector and is screwed anticlockwise; the oxygen hose fitting is screwed on clockwise.

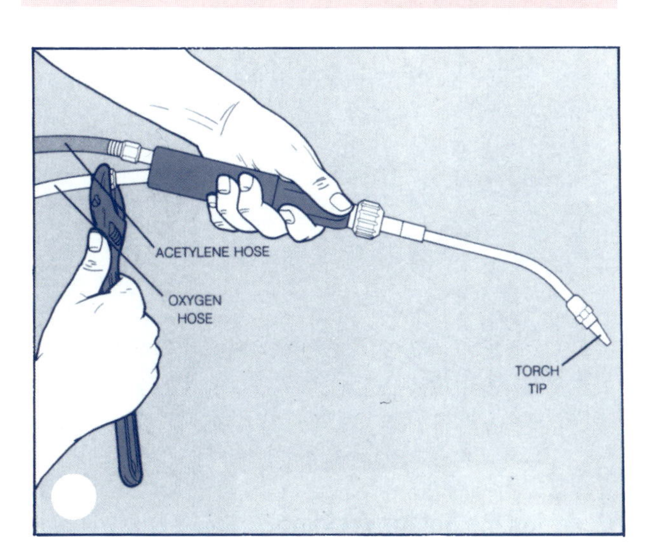

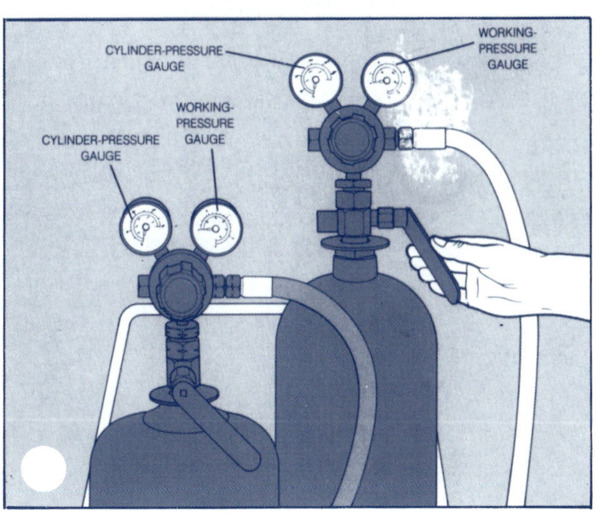

MIG, FCA or GMA?

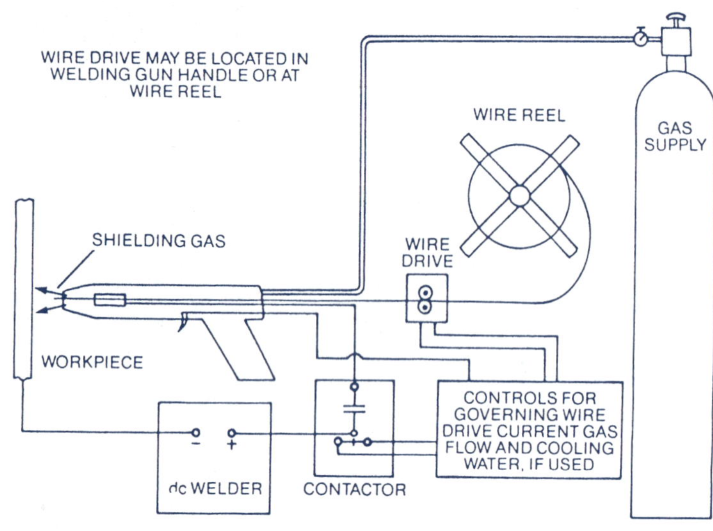

FIGURE 17-3 A schematic diagram of the gas-metal-arc welding process. (Courtesy Linde Division, Union Carbide Corp.)

For a MIG-welding unit you need:

- A suitable welding power source to provide current to melt the wire and base metal being joined.
- A wire feeder, contactor (switch) and controller to deliver the filler wire at the required speed.
- A welding gun to direct the wire and the shielding gas at the workpiece.
- A shielding-gas cylinder and high-pressure regulator with hose.
- A spool or coil of electrode wire of specified type and diameter for the work to be done—solid wire for MIG welding and cored wire for FCAW.
- Proper welding techniques for the given GMAW or FCAW process and base metal.

1 A basic set-up

Look at the schematic diagram and read the text.
First, underline the main parts of the welding set-up.
Discuss the words with a partner.
Then answer the following questions
by connecting the boxes.

What do these abbreviations stand for?

MIG ○	○ gas-metal arc
GMA ○	○ flux-cored arc
FCA ○	○ metal inert gas

What do you need these parts for?

power source ○	○ to deliver the filler wire at the right speed.
wire feeder ○	○ to provide current to melt the wire and these base metal.
welding gun ○	○ to direct the wire and the shielding gas at the work piece.
gas cylinder ○	○ to supply shielding gas to the welding gun.
solid wire ○	○ for FCA-welding.
cored wire ○	○ for MIG-welding.

Compare your results in groups of three.

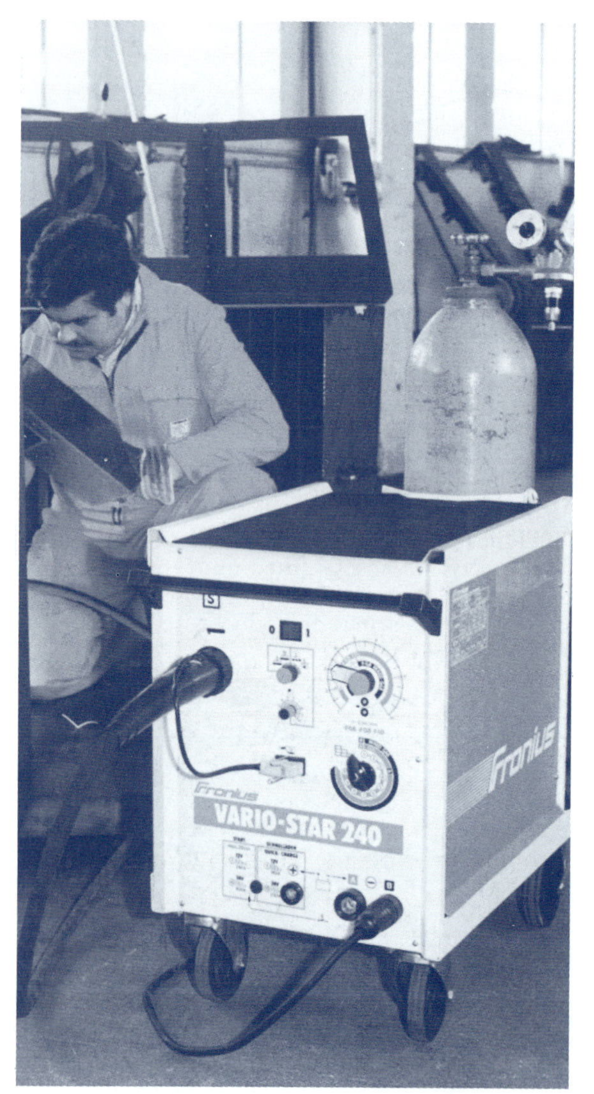

© Fronius

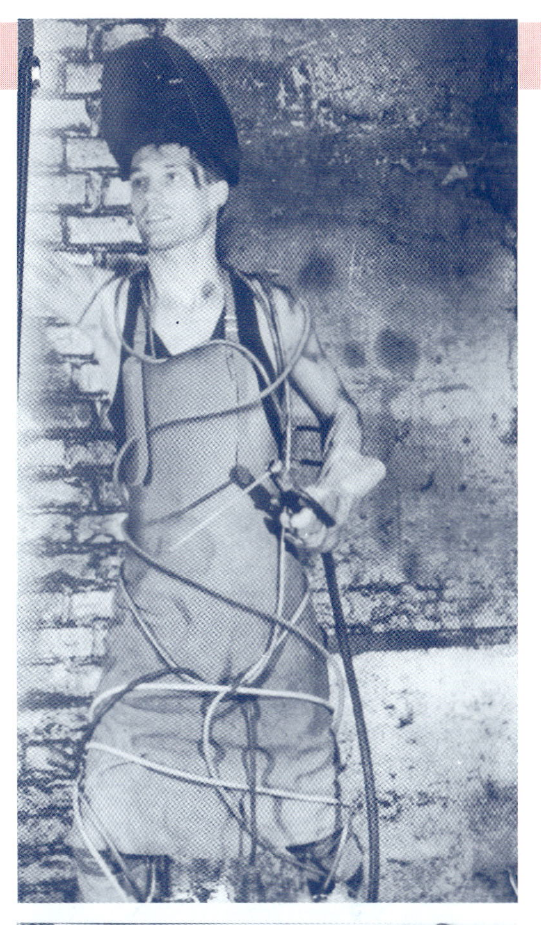

2 What is MIG-welding?

Look at the photos and listen to the interview on cassette.
What is it about?

....

Listen again and tick the answers to the following questions.

What is MIG-welding?
- ○ Welding MIG fighter planes.
- ○ Metal inert gas welding.
- ○ A wire-feed arc welding process.

What's the inert gas needed for?
- ○ To protect the arc.
- ○ To keep the oxygen away.
- ○ To protect the molten weld metal.

Compare your results with your neighbour.

3 Which gas?

Listen again:
Which gas do they use for which metal?
Connect the boxes.

stainless steel ○	○ Argon
copper ○	○ Oxygen
aluminium ○	○ Helium
steel ○	○ Ar-He mixture
	○ Carbon-dioxide
	○ Ar-CO_2 mixture

4 Which metals?

Listen once more and tick possible applications.

Where can you use MIG welding?

- ○ For maintenance and repair work.
- ○ In production processes.

For joining ○ sheet metal.
○ tubes and angles.
○ box sections.

For a wide range of metals, such as
○ low-carbon steel.
○ stainless steel.
○ aluminium.
○ magnesium alloys.
○ copper.

What do you know about MIG welding?
Discuss the process and its applications in groups of three.

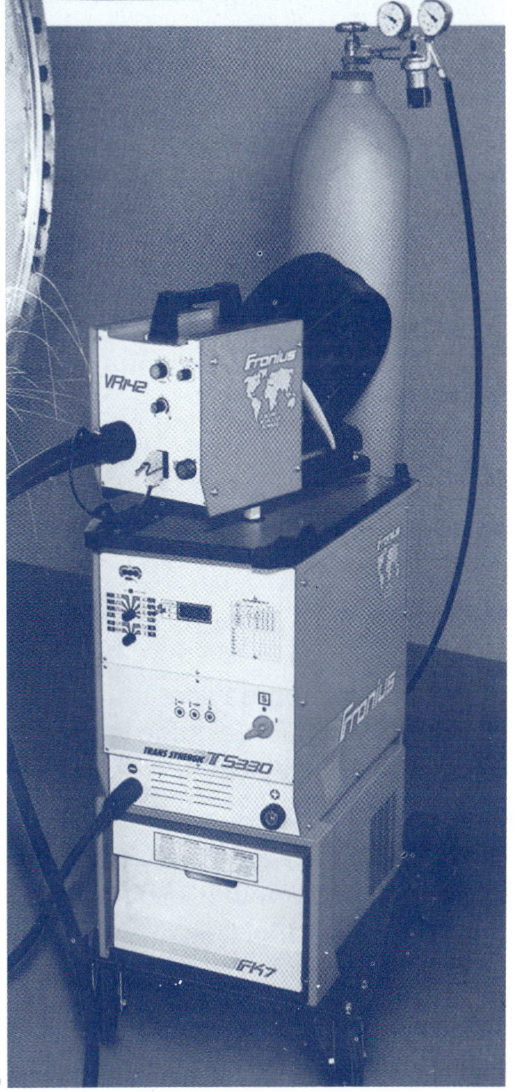

1 Number one

Look at the photo and listen to this scene.
Who or what is "number one"?

○ A welder.
○ A robot.
○ A computer.
○ A person from another planet.

2 Talk to the robot

Listen again and tick the answers to the following questions.

What is the robot working on?
○ A caterpillar.
○ An earth-moving device.
○ A bucket for the caterpillar.
○ A car body.

How is the workpiece moved?
○ By the robot.
○ By the rotation device.
○ By the longitudinal slide.
○ By the cross-slide.

What does the welder do?
○ Rotate the workpiece.
○ Select the computer programmes.
○ Talk to the robot.
○ Feed the robot with programmes.

Compare your results with a partner.
Then tell him or her what you have learned about "number one".

The RT 280

The precision robot RT 280 has 6 revolute joint axles in rigid aluminium cast construction. It is suited for upright and suspended assembly.
The main drive axle is designed as an extended angular machine foot bearing two rotating arms. The front arm turns around axle 3 and around its longitudinal axle (6). At the end it has a special joint with a hollow arbour, so that the welding torch can be moved around two further axles. This design provides for a vast operational range. A multiprocessor system guarantees precise path control. Its software is custom-made for welding applications and allows quick programming for all arc-welding tasks. A teach-in programme is available.

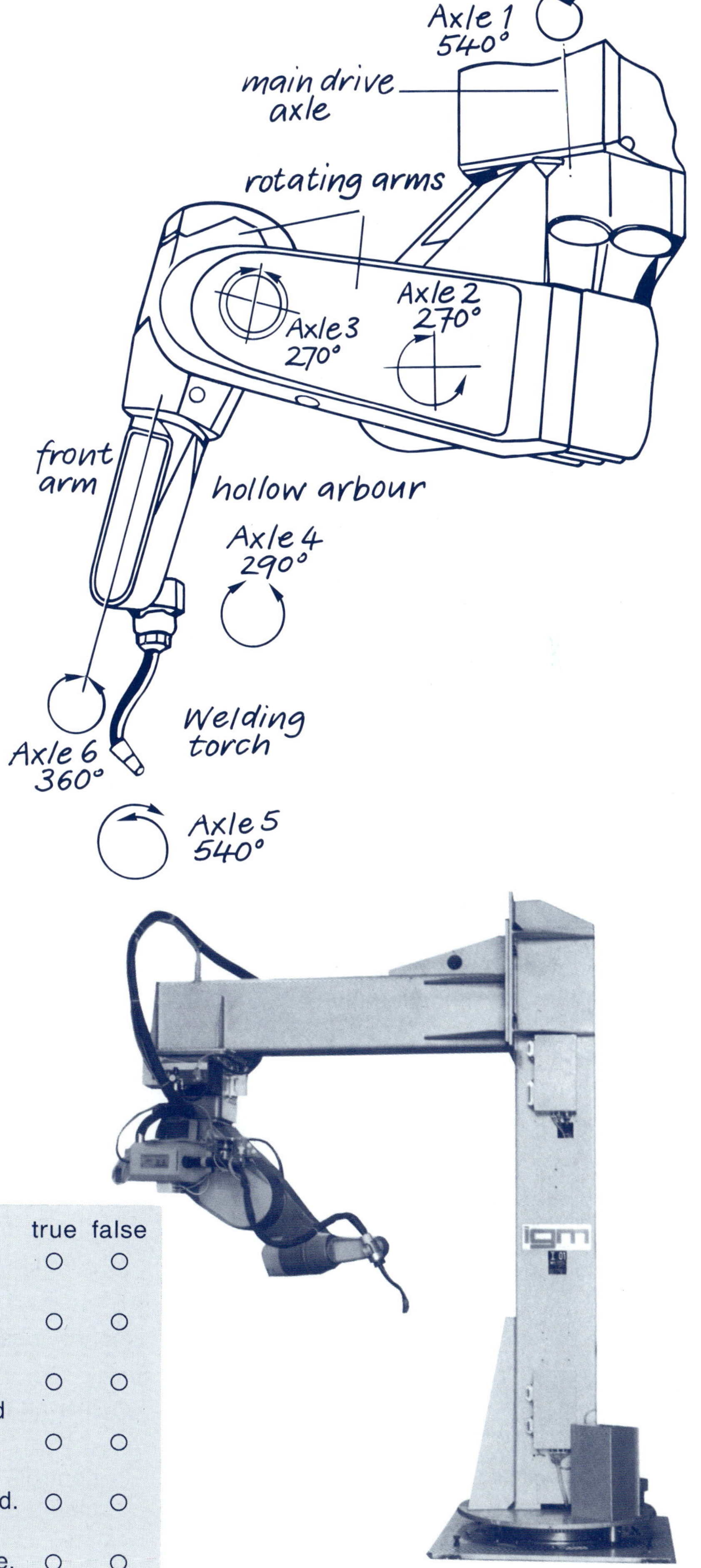

3 The RT 280

Look at the picture and read the text.
Underline all the parts which are necessary to move the welding torch.

Read the text again and mark the statements true or false.

	true	false
The RT 280 has 6 rigid joints.	○	○
It can be mounted upright only.	○	○
The front arm turns around its longitudinal axle only.	○	○
The welding torch moves around one single axle.	○	○
The computer-software is standard.	○	○
It includes a teach-in programme.	○	○

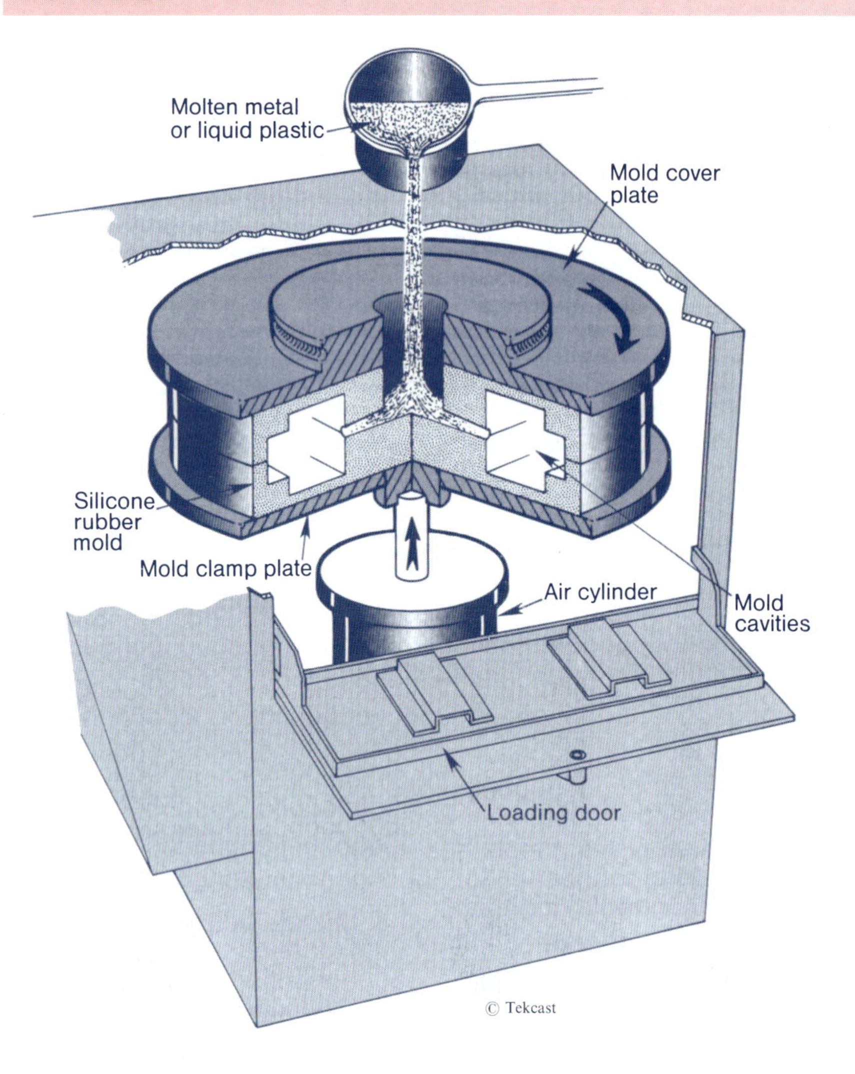

The unique, patented front-loading Tekcaster provides three key functions for the production of functional, highly engineered or detailed parts:

1 Mold stabilization without distortion. Tekcast's special, patented mold clamping system centers and supports the mold in a horizontal plane, while a pneumatic cylinder applies an even, closely controlled clamp pressure which holds the mold halves firmly together during pouring without distorting the mold cavity by over-squeezing.

2 Mold pouring under pressure. Rapid, speed-controlled rotation provides the adjustable centrifugal force which distributes the liquid metal or plastic throughout the mold, filling all cavities and intricate details *under pressure* before the material can solidify or set up.

3 Complete control of centrifugal force and pressure. Tekcaster is fully instrumented and automated to provide a complete range of spin speeds and clamp pressures. This optimizes parts quality, tolerances, precision and detail, while improving productivity of the complete spin-casting cycle.

1 The Tekcaster process

Look at the drawing and read the process description.
Underline all the words you know.
Look up three words in each paragraph and discuss them with a partner.
Connect the following statements.

The mold clamping system ○	○ applies clamp pressure to hold mold halves together.
A pneumatic cylinder ○	○ centres and supports the mold horizontally.
Rapid rotation ○	○ distributes liquid metal or plastic throughout the mold.
Centrifugal force ○	○ is fully instrumented and automated.
Liquid metal or plastic ○	○ provides centrifugal force.
Tekcaster ○	○ fills all cavities and details.
	○ controls spin speed and clamp pressure.

Compare your result in groups of three.

2 The silicon rubber mold

Listen to the telephone call.

What are they talking about?

- ○ Plastic injection molding
- ○ Pressure die casting
- ○ Spin-casting

Listen again and tick the right answers.

What does the customer's company produce?

- ○ Small plastic parts
- ○ Cog-wheels
- ○ Lawn-mowers

Why is he interested in spin-casting?

Because for small quantities

- ○ spin-casting is more cost-effective.
- ○ spin-casting is cheaper.
- ○ plastic injection molds are very expensive.

Compare your results with a partner.

3 Sounds good

Listen to the phone call again and answer the following questions.

How much would a mold for six cog-wheels cost?

About ....

How long would a silicone rubber mold last?

About .... casts.

How long does it take to run off a cast?

- ○ A few days.
- ○ A few hours.
- ○ Several weeks.

PLACE STAMP HERE

Tekcast Industries, Inc.
P.O. Box 677
New Rochelle, N.Y. 10802, U.S.A.

4 Interested?

Would you like to know more about the Tekcast process? Fill in the postcard and write a polite letter saying that you would like to visit them with a group of students.

IW 6/89 Date________

WE ARE INTERESTED AND WOULD LIKE TO KNOW MORE:
(Please complete, print or type)

- □ We would like to make arrangements to visit TEKCAST.
- □ We would like to send in part(s) to be Spin-Cast.
- □ We are interested in Spin-Casting parts in □ Metal □ Plastic
- □ Call me, I have some questions.
- □ We would like to receive additional technical information.
- □ Send us equip. specs. & price quote. System size: TEK- ________

Our major products:________________________________

Potential application(s): ____________________ Part size(s)____________

Name ______________________________ Title____________

Company ______________________________ Dept.____________

Mailing Address ______________________________

City ______________________________ Country____________

Phone: ( ) ____________ Telex: ____________ Fax: ( ) ____________

My interest is: □ Immediate □ 3-6 months □ Information Only.

□ Our only interest is to purchase Spin-Castings

We are: □ Potential User, □ Distributor

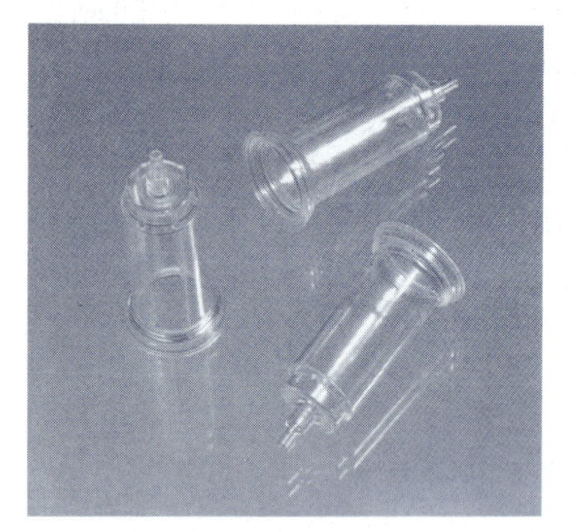

Intravenous infusion chambers – lower parts

32-cavity injection mould for manufacturing ....

....

Shaver housing of 2 components –
Housing: ABS
Non-slip grip surface: Elastomer

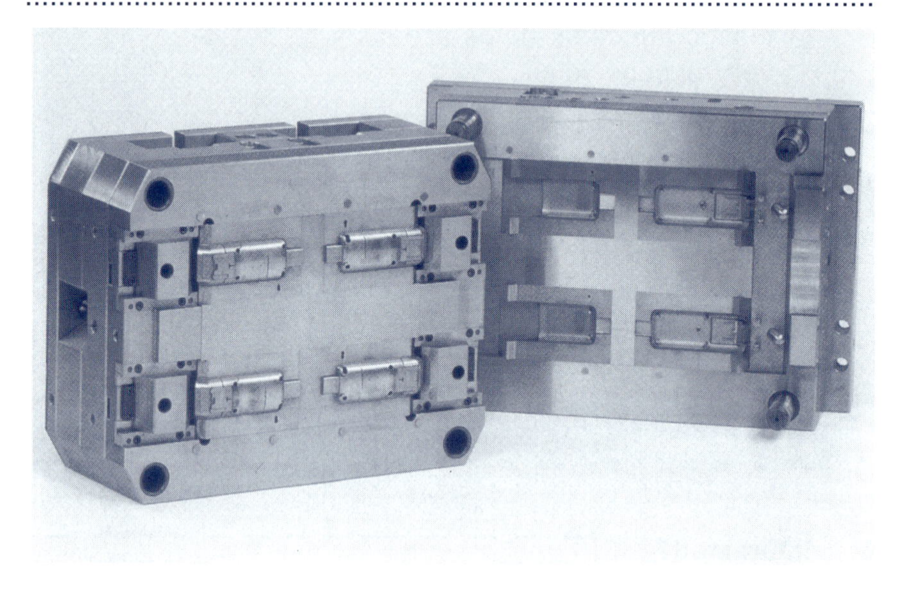

2-component injection mould for manufacturing ....

....

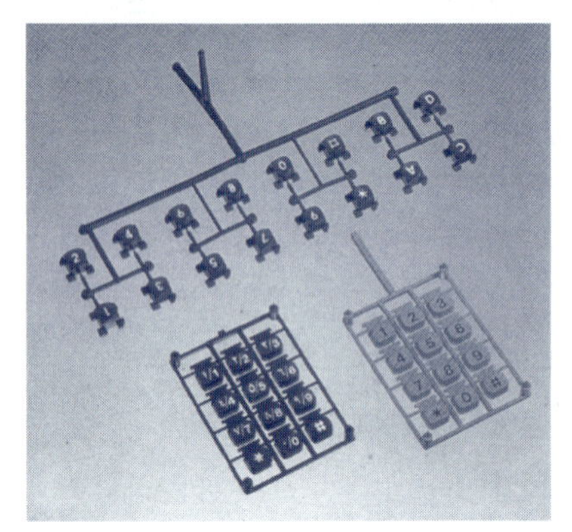

Telephone keys

1 Infusion chambers

Look at the photos on this page.
Which of the products can be made in which mould?
Complete the text below the photos.

What are the products made of?
○ Rubber.
○ Thermoplastic.
○ Plexiglass.

Where do you think these moulds are being used?
○ In machining centres.
○ In injection moulding machines.
○ In spin-casting machines.

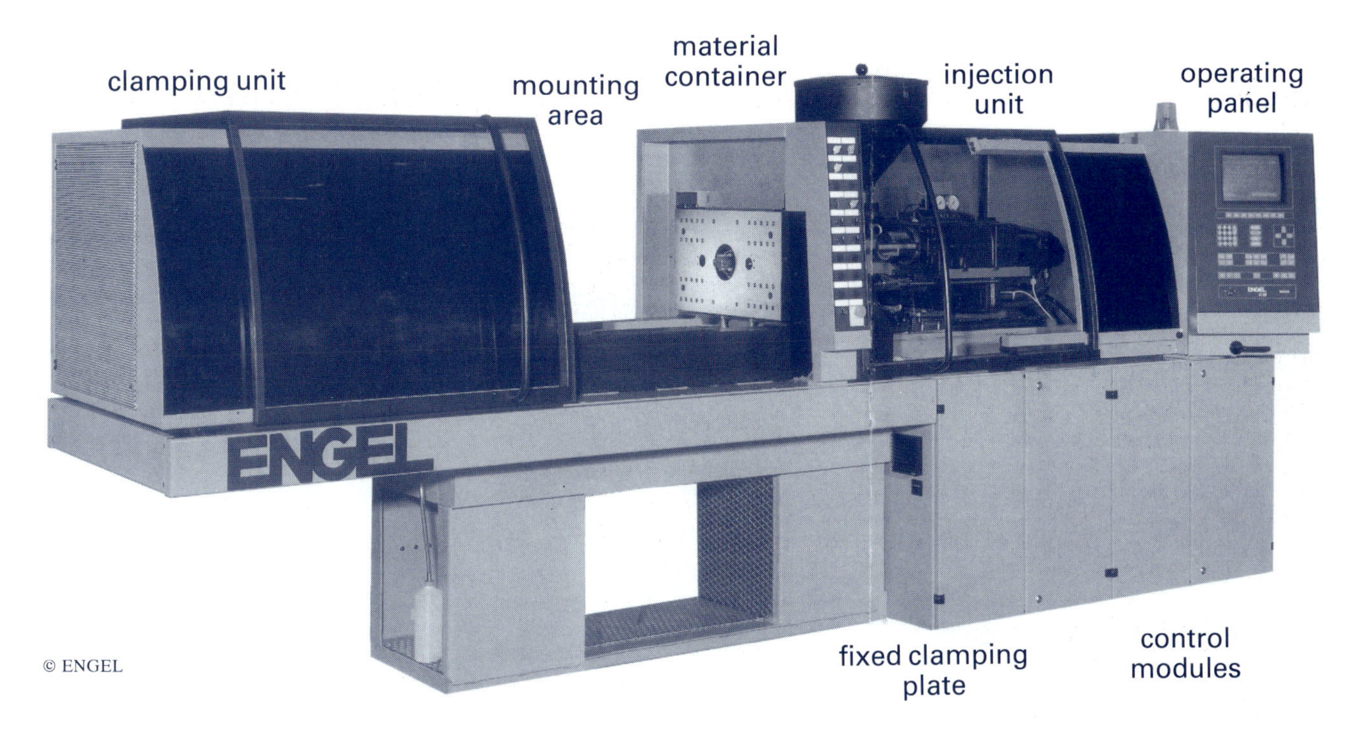

The heart of any injection moulding machine is the injection unit. Raw material e.g. granulated thermoplastic is fed into the material container. A hydraulic system drives the screw which plasticises and transports the mass.

Then the mass in injected into the mould at maximum speed and pressure. The mould is held in place in the mounting area by a clamping unit. All the units of the injection moulding machine are centrally controlled from the operating panel which is connected to a central computer system.

2 Machine units

Look at the photo of the injection moulding machine and read the description.
Underline the machine units in the text.
Read the description again and fill in the gaps.
Use the following verbs:

connect
control
drive feed
hold inject
plasticise.

Change the verbs to fit them into the sentences.

Granulated thermoplastic	into the material container
and	by the screw.
The screw	by the hydraulic system.
The mass	into the mould.
The mould	in place by a clamping system.
All the units	from the operating panel.
The operating panel	to a central computer.

3 Changing the moulds

Look at the photos on the opposite page and listen to the conversation on the shop floor.
What's the machine operater doing?

....

Which parts are being made?

....

Listen again and complete what the operater says:

The mould has two parts.
Yes, they're mounted ....
Now
like this.
And now shut her tight with ....
Here we are. Ready ....

Tool changing, fourth axis and pallet systems can all be specified on the Kearns-Richards range of heavy duty machining centres. They can be equipped with CNC contouring facing heads with 100, 130 or 160 mm diameter spindles.

K-R - the headmasters

CNC FACING HEAD SOLVES PROFILE GRINDING PROBLEMS

A CNC head machining a joint.

If there is one thing for which Kearns-Richards is renowned, it is producing machines with integral CNC contouring or facing heads. For further flexibility the CNC facing head can be equipped with a grinding spindle to adapt the machine for internal and external planetary grinding.

The overall benefits of CNC facing heads can be dramatic, but it needs an appreciation of the potential presented by the techniques before such benefits can be fully realised.

For those unfamiliar with the technique, the contouring head has two main features. First there is a central spindle that uses conventional rotating tools. The unique feature is the slide built into the head that can be positioned radially under full CNC control. This head is independently mounted from the spindle and incorporates its own drive system.

The CNC facing slide, which moves radially, is driven through a ballscrew and nut assembly via an anti- backlash differential gear assembly.

Using the U axis as the radial axis is designated in conjunction with the 'W' axis (table feed towards the column) and rotation of the head, it is possible to generate an internal or external spherical profile. Similarly tapered and stepped bores as well as threaded components can be produced with a single tool.

All too often such component details are produced by a subsequent operation on a vertical lathe. The contouring head reduces the number of settings and ensures critical concentricities are maintained by completing bores and profiles in one setting.

A recent application that is attracting a lot of attention for the planetary grinding technique is the refurbishment of aircraft undercarriage legs. Here the component is held stationary whilst the planetary grinding head accurately grinds the outer diameter whilst maintaining concentricity with other component features.

The slide of the CNC contouring head.

1 K-R CNC

Look through the opposite page.
What sort of text is it?

○ A data sheet.
○ Part of a leaflet.
○ Instructions for use.

2 Facing and contouring

Read the text again and tick the right statements.
K-R produce

○ CNC machining centres.
○ CNC contouring heads.
○ CNC facing heads.
○ CNC grinding spindles.

The grinding spindle can be used

○ for facing.
○ for internal grinding.
○ for external grinding.
○ for contouring.

The contouring head

○ has a central spindle.
○ has a radial slide.
○ can take a grinding spindle.
○ can take conventional rotating tools.

The CNC facing slide

○ has its own drive system.
○ moves radially.
○ moves along the U-axis.
○ is driven by a ballscrew and nut assembly.

Compare your result with a partner.

3 Bores and profiles

Look at the drawings and connect them to the following terms.

stepped bore ○

tapered bore ○

external spherical profile ○

internal spherical profile ○

threaded components ○

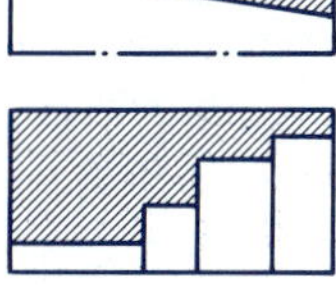

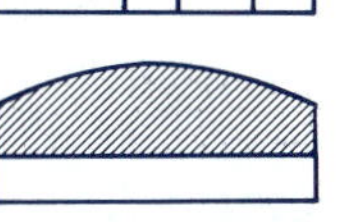

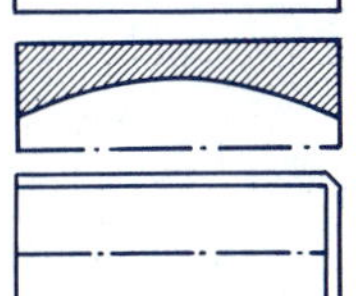

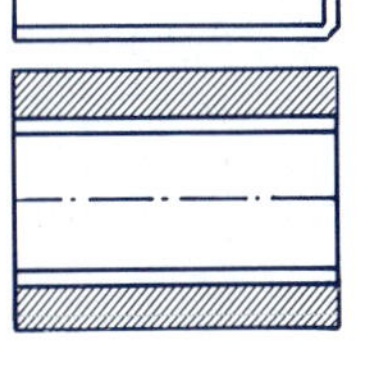

What is the text about?
About ○ CNC machines.
○ your headmaster.
○ CNC facing and contouring.

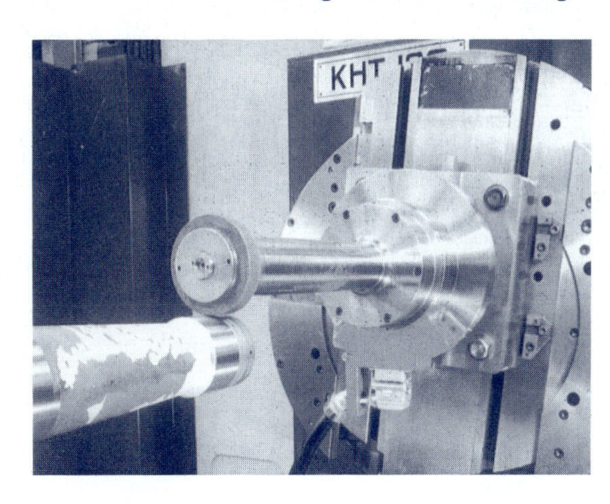

4 Planetary grinding

Look at the drawings and listen to the conversation.
Which drawing are they talking about?

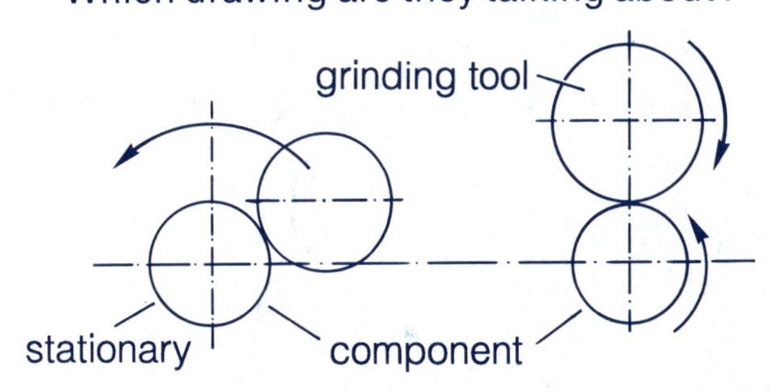

Which of the processes do they call conventional grinding?
Which is planetary grinding?

Listen again and connect the following statements.

Fill in Ⓒ for conventional
and Ⓟ for planetary grinding.

◯ The tool is rotating.
◯ The component is rotating.
◯ The component is stationary.
◯ The tool is moved longitudinally.
◯ The tool is moved around the component.

Tell a partner about conventional and planetary grinding.
Can you describe other machining processes?

MSA understands the hazards you face better than any other supplier of occupational safety and health equipment in the world. That's why we manufacture over 4,000 products for your protection. All backed by over 75 years of proven, on-the-job experience.

So no matter where you are... no matter what you do... MSA is there with the quality personal protection you can use with confidence day after day. Products that provide the specific protection that meets your job needs.

Respiratory Protection: A full line of NIOSH/MSHA-approved respirators for protection against a wide range of toxic gases, vapors, dusts, mists, fumes and radionuclides.

Instruments: All designed to accurately detect, indicate, signal, measure, analyze, monitor and warn against the threat of toxic and combustible gases and oxygen deficiency. Hand-held, portable or permanently installed.

Head Protection: Hats and caps for industrial workers, welders, electrical and construction workers, miners and others.

Hearing Protection: A complete line of ear plugs and ear muffs protect against noises in factories and airports, and where jackhammers, air compressors or turbines are used.

Eye and Face Protection: Over 20 different types of goggles—industrial, chemical splash, welders, chippers, plant visitor, etc. Also faceshields and accessories.

Body Protection: MSA makes special belts and harnesses for a variety of specific applications. Also chemical-protective, flame-retardant and heat-resistant garments; air-cooled suits; anti-contamination garments; and gloves for virtually every industry.

Discover the security that MSA quality and reliability can bring to your operation.

For more information and a copy of our "MiniGuide to Safety Equipment," call or write one of our 22 worldwide subsidiaries today. Or one of over 100 MSA representatives around the world.

MSA International, P.O. Box 426, Pittsburgh, PA 15230 USA. Or phone (412) 967-3256, Telex 812453 or Telefax (412) 967-3451.

Count on MSA.

MSA

PLEASE USE YOUR SAFETY GEAR

1 Count on MSA

Look at the text on the opposite page.
What kind of text is it?

○ Process instructions
○ Advertisement
○ Product information

What kind of company is it about?

A manufacturer of ○ space fashion
○ safety equipment
○ deep-sea suits

2 Hazards you face

Read the text carefully and connect the boxes.

What kind of products do they advertise?		What are the products good for?
respirators ○		○ protect industrial workers' heads.
instruments ○		○ protect you against toxic gases, fumes, dust etc.
hard hats ○		○ warn against toxic and combustible gases.
ear plugs ○		○ protect you against noise at work.
ear muffs ○		○ protect your face.
face shields ○		○ protect your eyes.
special suits ○		○ protect your feet.
gloves ○		○ protect your hands.
safety shoes ○		○ protect you against chemicals, fire etc.

Test a partner like this:

To protect your head.

And what's an air-cooled suit for?

To protect you against heat.

3 Safety gear

Listen to Jack at the steel mill.
What's his problem?

....

Listen again and tick the garments above that Jack will have to wear.

What are you required to wear at work?
And at school?

Unit 32 Safe on two wheels

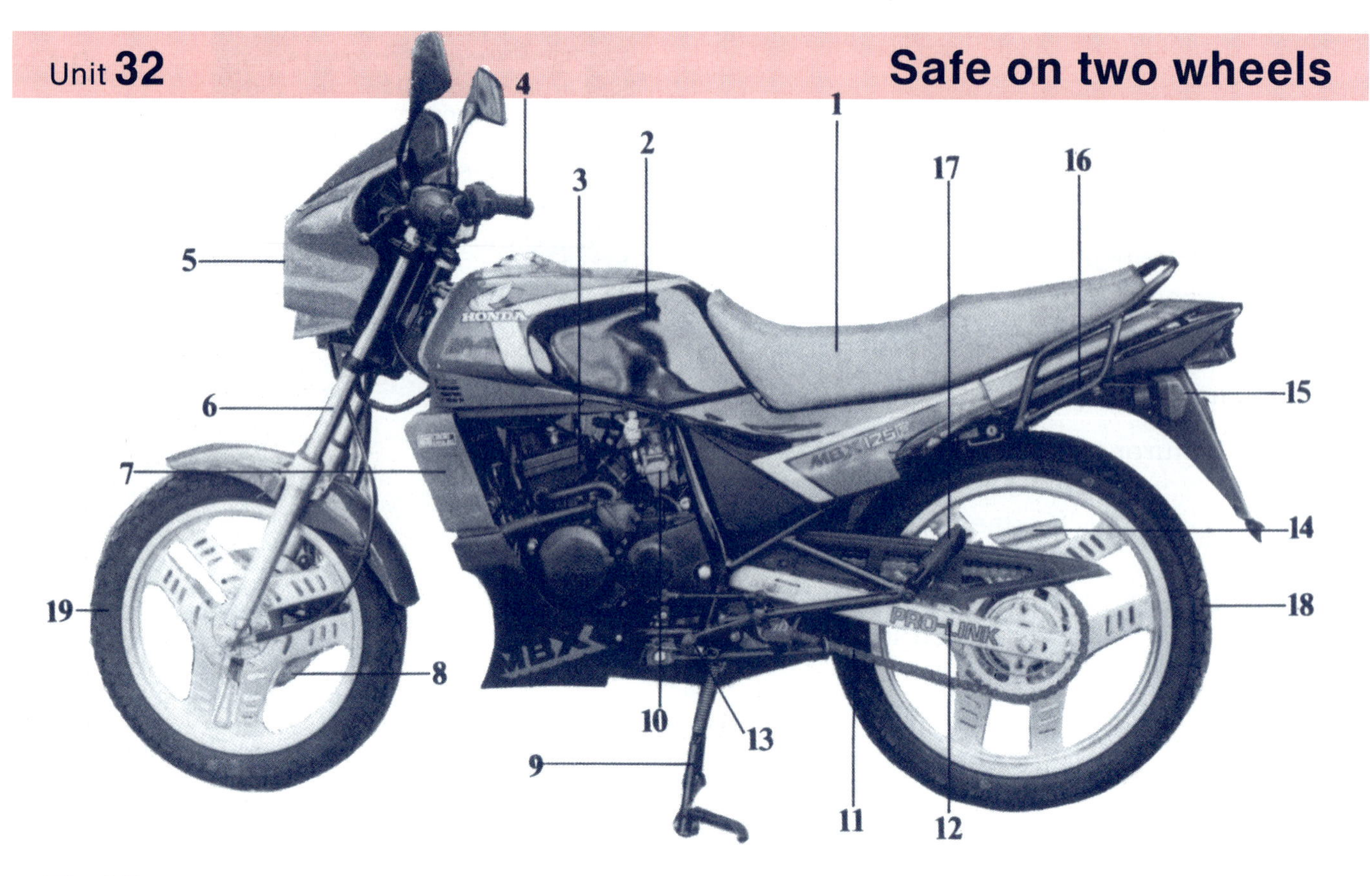

The bike

The type of bike you buy will depend on two things – your age, and how much money you can afford to spend. If you are 16 you may only ride a moped.
Prices range from £ 250 to £ 750 for a new one. £ 100 up for a second-hand one.
If you are 17 or older and a learner you may ride a bike of up to and including 125 cc.
Prices range from £ 500 to £ 1100 for a new one. £ 300 up for a reasonable second hand one.
If you are 17 or older and hold a full licence you may ride a bike of any cc. Prices are from £ 600 up for a reasonable second hand 500 cc machine.

1 Parts of a bike

Look at the picture and read the list of parts.
Number them in the list.

- ◯ seat
- ◯ petrol tank
- ◯ engine
- ◯ handlebars
- ◯ headlight
- ◯ front forks
- ◯ radiator
- ◯ front disc brake
- ◯ gear lever
- ◯ silencer
- ◯ prop stand
- ◯ carburettor
- ◯ chain
- ◯ tyres
- ◯ rear foot rest
- ◯ passenger grab handle
- ◯ rear suspension frame

2 Moped or bike?

Read the text on the left and fill in the following grid.

		minimum age		price range	
		GB	D	GB	D
Moped up to 50 cc	new second hand				
Motor bike up to 125 cc	new second hand				
Motor bike unlimited cc	new second hand				

Compare your results in groups of 3 or 4, like this:

*In GB you may ride a moped from the age of 16.
A new moped will cost from 500 pounds up.*

Can you re-write the text for Germany?

Checklist

Look for
- ○ worn tyres
- ○ loose spokes
- ○ frayed cables
- ○ oil leaks
- ○ loose silencer
- ○ dry or worn chain
- ○ any scratches
- ○ bent frame or footrest
- ○ light controls: brake lights, headlight, indicators

Test
- ○ the brakes
- ○ the horn
- ○ the headlight
- ○ the brake light
- ○ the indicators
- ○ the wheel bearing
- ○ the swingarm
- ○ the spokes
- ○ the chain by lifting it off the sprocket

© KTM

3 Chain's a bit dry

Look at the checklist and listen to Tina at the bike dealer's. Tick the items they discuss.

Listen again and mark the items
⊕ if they are in order, or
⊖ if they need adjustment, repair or replacement.

The chain The rear wheel bearing The front brake The rear tyre	needs to be should be	greased tightened adjusted replaced
Frayed cables Oil leaks Loose spokes Controls	need to be should be	fixed repaired sealed oiled

Listen to Tina again and complete the bubbles.

Look at the chart and compare your results with a partner.

Clutch and front brake feel ....

The brake light is ....

4 A second-hand bike

Think of a second-hand bike you would like to buy.
What do you say in case things are wrong with the bike?
Your partner can be the bike dealer.

Dependable as a horse

"ALLWORK"

Light Tractor

Pulls Three Plows Easily

A Four-Wheel Tractor Dependable as a Horse

A simple, durable, powerful machine selling at a price you can easily afford. Equipped with four-cylinder vertical engine 5x6, developing 25 h. p. at belt, 12 h. p. at drawbar. Two-speed transmission working in oil, automobile type front axle, roller-bearing rear axle, steel gears thoroughly protected from dust and self-oiling, radiator and fan that cool absolutely, 16-inch face rear wheels, weight 4800 pounds.

A sensible, practical tractor, built by a company with an established reputation and numerous machines at work in fields today. Write for catalog.

Electric Wheel Co. Box 50A, Quincy, Ill.

ENGINE	Steyr 8120
Output kW/DIN-hp/SAE-hp	73.5/100/110 with turbocharger
Output at rev/min	2300
Bore/stroke (mm/mm)	100/110
Cylinder capacity (cm³)	5182
number of cylinders	6
Max. torque (Nm/kpm) at rev/min	370/37.7/1400
Torque increase (%)	21
Water-cooling	long-life anti-freeze
Tank capacity (litres)	122

1 Allwork

Look at the advertisement for a light tractor and answer the following questions.

What do they say about the machine?

○ simple ○ powerful
○ durable ○ inexpensive

What do they say about the engine?

Number of cylinders: ....

Bore × stroke: .... inches

Power at belt/drawbar: ....

What do they say about the following parts:

Transmission: ....

Front axle: ....

Rear axle: ....

Rear wheels: ....

How much does the tractor weigh? ....

h.p. = hp . . . horsepower: 1 hp = 0.75 kW

Compare your results with a partner. Then compare the data of the "Allwork" with the data for the "8120".

2 It's got 100 hp

Look at the engine data of the old and new tractor on this page. Listen to the conversation and tick the data they are talking about.

Listen again and mark the following statements

Ⓐ for "Allwork" and
Ⓢ for the "8120"

and fill in the missing data.

The ◯ was built around ....

◯ is a lot more powerful.

◯ has got . . . hp.

◯ has got . . . hp at the drawbar.

◯ is lighter than the ◯

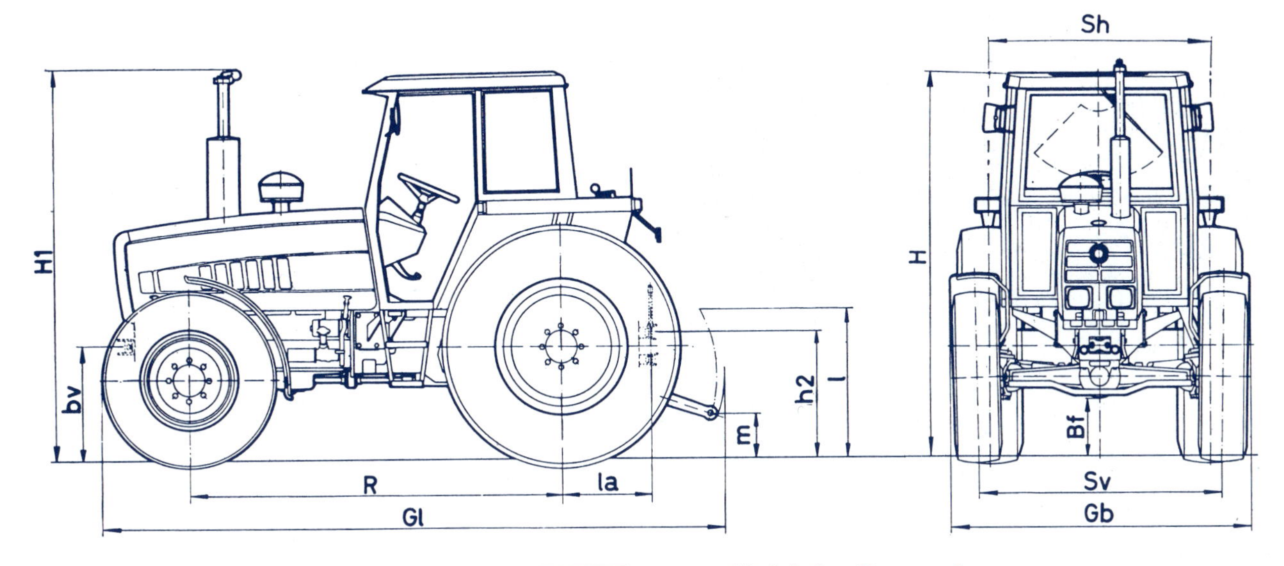

VEHICLE DIMENSIONS (in mm)

	8120
Gl. = max. length	4150
Gb. = max. width	2140
H = max. height with driver's protection	2595
H1 = max. height up to exhaust pipe	2626
R = wheel base	2509
la = center rear axle to center trailer coupling	668
Bf = ground clearance of front axle	440
bv = height of trailer coupling, front	760
Sv = track width, front	1478–1878
m = deepest position of connection point at basic adjustment	320
h2 = height of trailer coupling center, adjustable	884–406
l = max. lifting height over ground	1000
Sh = track width, rear	1654

VEHICLE WEIGHTS (in kgs)

Service weight of tractor	3980
Front axle load	1280
Rear axle load	2700
Permissible total weight	6000
Permissible rear axle load *	5000
Permissible front axle load *	1750/1800
With wheel load	1500
Ballast weights, front	330–595
Ballast weights, rear	500

*) Within the range of the permissible total weight

DRIVING SPEEDS (in km/h) at full-load engine speed of 2,300 r.p.m.

TYPE 25 km/h (12/4-gear transmission)
TYPE 30 km/h (15/5-gear transmission – optional *)

	L	M	S	R	KR **
1st gear	2,03	4,11	8,62	5,04	0,38
2nd gear	3,10	6,27	13,16	7,70	0,58
3rd gear	3,80	7,69	16,13	9,44	0,71
4th gear	5,50	11,15	23,39	13,69	1,03
5th gear *	6,75 *	13,69 *	28,70 *	16,80 *	1,26 *

L = slow range: for heavy ground work
M = medium range: for medium ground work
S = fast range: for transport
R = reverse driving range
KR = crawling range (optional **)

3 Vehicle dimensions

Look at the drawing of the 8120.
Study the Vehicle Dimensions.
Help a partner to transfer the measurements onto the drawings.

H1 is the maximum height up to the exhaust pipe.

It's two thousand six hundred and twenty-six millimetres.

Then ask your partner to help you.

4 How much can you load?

Ask your partner the following questions:
How much does the tractor weigh?
What's the service weight of the tractor?
What's the load on the front axle?
And on the rear axle?
What's the total, permissible load on the rear axle? Etc.
You can choose from the following answers.

1500 kg	fifteen hundred kilos. One thousand five hundred kilos.
3980 kg	Three thousand nine hundred and eighty kilos. Tree point ninety-eight tons.
6000 kg	Six metric tons. Six thousand kilos.

5 How fast can it go?

Study the driving speeds in kilometres per hour at twenty-three hundred r.p.m. (revolutions per minute).
Talk to a partner about it like this:

In the fast range, for transport. the 8120 can go .... km/h in the fourth gear.

Controlling pollution?

Waste gases
Modern engines are modified internally and are equipped with many add-on devices to reduce the formation or prevent the escape of three chemicals that are hazardous to human health:
Hydrocarbons (HC) are unburned or partly burned gasoline. HC may evaporate from the crankcase or fuel system, or may be found in exhaust gases due to incomplete combustion.
Carbon monoxide (CO) is formed when fuel combustion is incomplete because too little oxygen is available (the fuel mixture is too rich).
Oxides of nitrogen (NOx) are formed in the combustion chamber when peak temperatures are high enough to force the oxygen and nitrogen in the air to combine. HC and NOx react with sunlight to form smog.
Controlling all three of these pollutants at once is difficult. Where possible, internal modifications have been made to the engine. Compression ratios have been lowered to cut peak temperatures. Combustion chambers have been redesigned to eliminate little nooks. Camshafts have been designed to close the exhaust valve early. An external exhaust gas recirculation system is used to dilute the fuel charge.

1 Waste gases

Read the text and underline the chemicals you can find.

Then read the statements and connect the boxes.

Hydrocarbons ○	○ may evaporate from the crankcase.
	○ may be found in exhaust gases.
	○ are unburned or partly burned gasoline.
	○ is formed when the fuel mixture is too rich.
Carbon monoxide ○	○ is formed when fuel combustion is incomplete.
	○ react with sunlight to form smog.
	○ is formed when not enough oxygen is available.
Nitrogen oxides ○	○ are formed in the combustion chamber.

2 Controlling pollutants

Read the text again:
Which modifications can be made to control pollutants?

....

....

3 Catalytic converters

Look at the drawing and listen to the interview.
Tick the words in the drawing when they are mentioned in the interview.

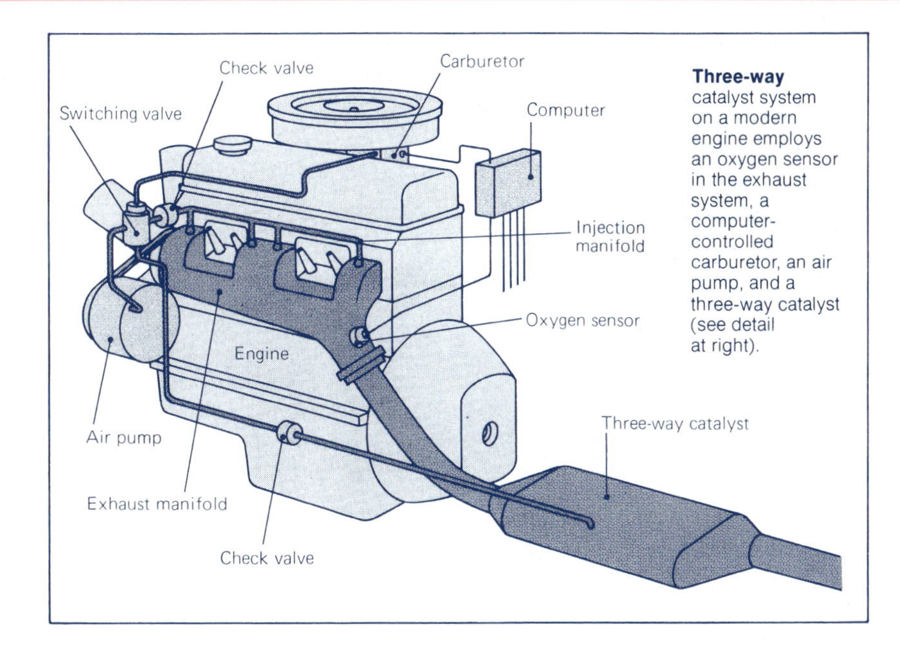

Three-way catalyst system on a modern engine employs an oxygen sensor in the exhaust system, a computer-controlled carburetor, an air pump, and a three-way catalyst (see detail at right).

Listen again and tick the right statements.

The catalyst turns exhaust emissions into
- ○ harmless substances.
- ○ water.
- ○ relatively harmless substances.

Catalytic converters consist of
- ○ two sections.
- ○ a three-way catalyst.
- ○ an oxidation catalyst.

Catalytic converters
- ○ change exhaust emissions.
- ○ reduce exhaust emissions to zero.
- ○ reduce exhaust emissions under certain conditions.

The greenhouse effect is caused by
- ○ carbon dioxide.
- ○ hydrocarbons.
- ○ nitrogen oxides.

Catalytic converters work under the following conditions:
- ○ Unleaded fuel must be used.
- ○ The air-fuel mixture must be 14.6 : 1.
- ○ The carburettor must be computer-controlled.

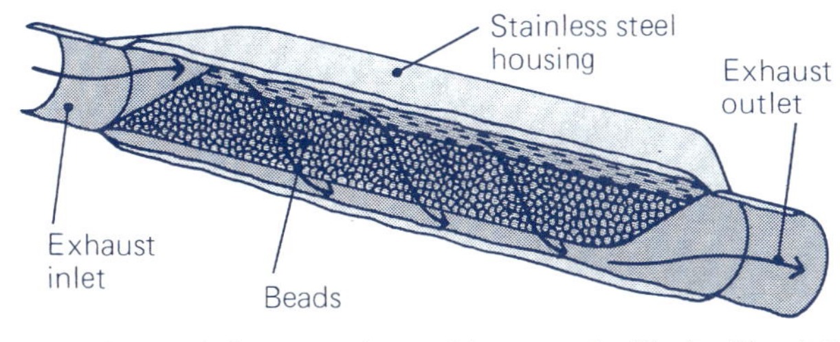

Oxidation catalyst used on older cars is filled with platinum-coated ceramic beads that reduce HC and CO to H_2O and CO_2. Some versions use a platinum-coated honeycomb instead of beads.

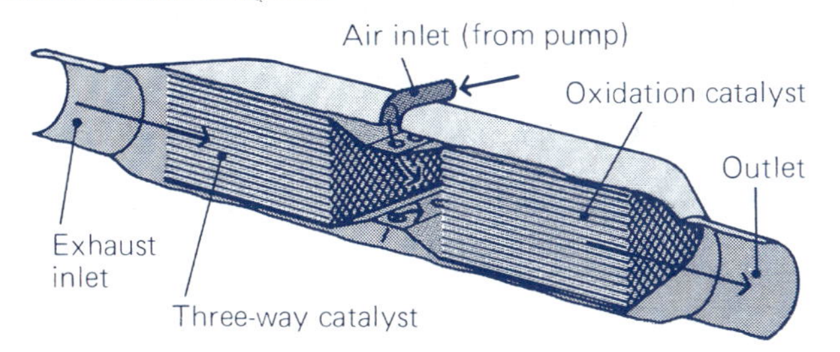

Three-way catalyst has a honeycomb coated with three rare metals: platinum, palladium, and rhodium. The rhodium reduces NO_x when the air-fuel ratio is just right. When the air-fuel ratio is not 14.6:1, the oxidation catalyst continues to reduce emissions of HC and CO.

4 Oxidation and three-way catalyst

Look at the drawings and read the paragraphs.
Connect the statements

Oxidation catalysts ○	○ are used in old cars.
	○ use platinum, palladium and rhodium.
Three-way catalysts ○	○ use platinum-coated ceramic beads.
	○ use a platinum-coated honeycomb.
Ceramic beads ○	○ reduces NO_X.
Rhodium ○	○ reduce HC and CO to H_2O.

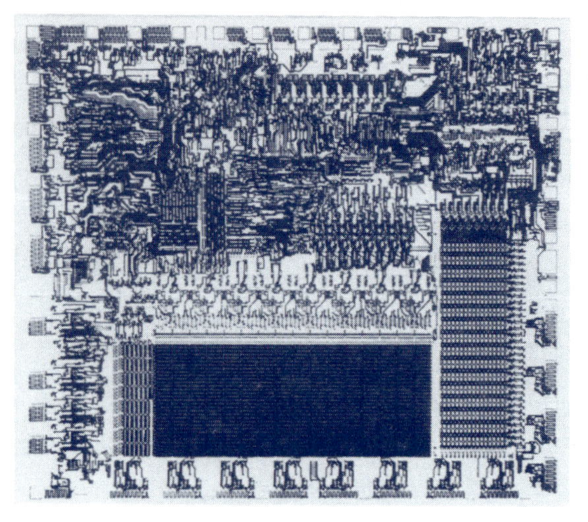

Minicomputers called *microprocessors* are used to control many systems in modern cars, including spark timing, airfuel mixture, engine idle speed, automatic choke settings, emission controls, and automatic transmission lockup clutches.
The computer box has many circuits and parts, but the computer itself is a tiny chip, programmed to make calculations based on information from various sensing devices. It then issues commands, which do nothing more than supply or deny current to certain electrical terminals to which the chip is connected. These terminals are connected to the operating hardware, which includes electric motors and solenoids.
The computer's sensors may monitor engine coolant and air temperatures, throttle and crankshaft positions, atmospheric pressure, manifold vacuum, pinging, engine speed, the concentration of oxygen in the exhaust, and car speed.

1 Minicomputers

Read the text above.
Which systems in a car can be controlled by a microprocessor?

....
....
....
....
....

What can be monitored by sensors?

....
....
....
....
....

Compare your results with a partner, like this:

Spark timing can be controlled by a microcomputer.

The engine codant can be monitored by a sensor.

2 How does it work?

Study the text above and connect the following statements.

The sensors ○	○ receives information from the sensors.
	○ monitor certain engine values.
The computer ○	○ are connected to the chip.
	○ issues commands to electrical terminals.
The terminals ○	○ includes electric motors and solenoids.
	○ are connected to the hardware.
The operating hardware ○	○ controls certain engine systems.

3 Hardware and sensors

Look at the schematic drawings on the opposite page and read through the descriptions of sensors and operating hardware. Connect the components to the engine where they are fitted.

Operating hardware

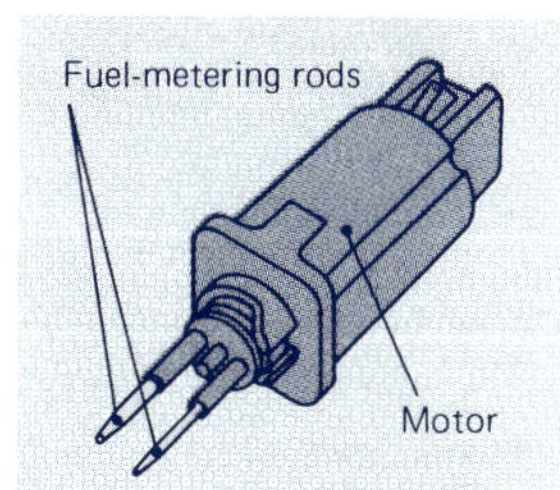

Motor on carburetor closes or opens a fuel-metering port to regulate fuel mixture, keeping the air-fuel ratio at exactly 14.6:1 (stoichiometric) at all times.

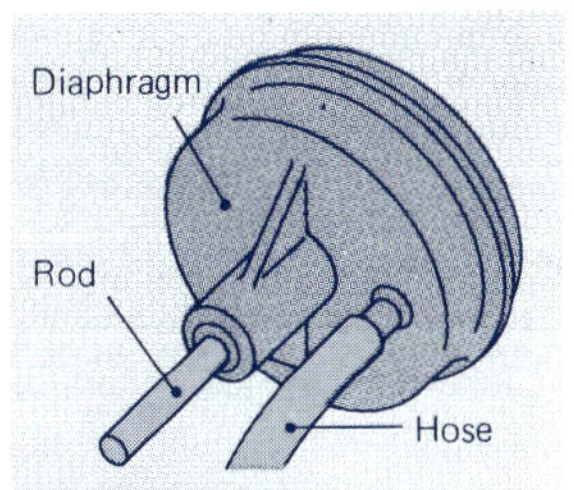

Throttle kicker uses a vacuum diaphragm to push a rod against the throttle linkage to increase idle speed. A solenoid valve controls vacuum supply.

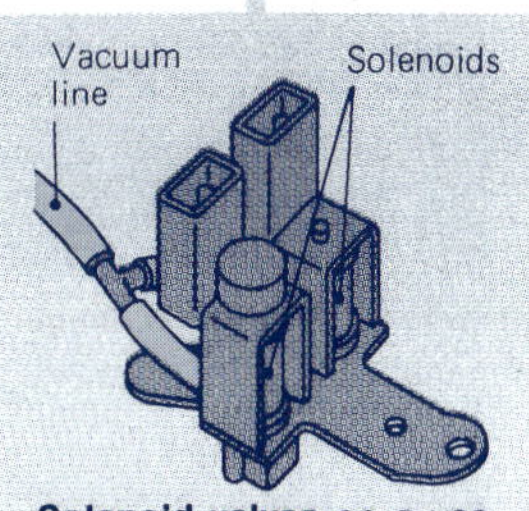

Solenoid valves on a vacuum line and a vent line can be operated by the computer to modulate the action of an engine's vacuum-controlled EGR valve.

Sensors

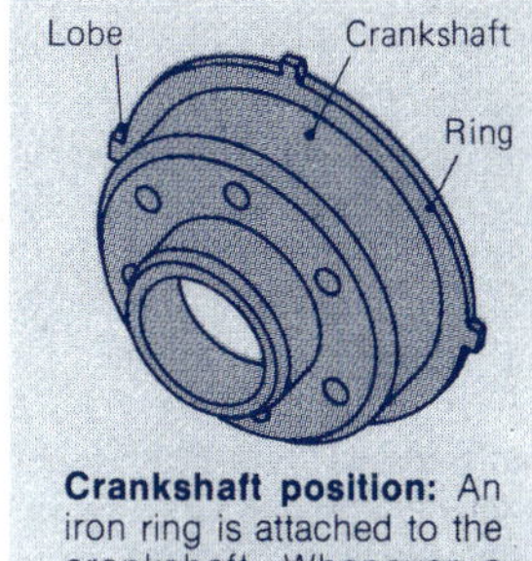

Crankshaft position: An iron ring is attached to the crankshaft. Whenever a lobe on the ring passes the magnetic sensor, a signal is sent to the computer.

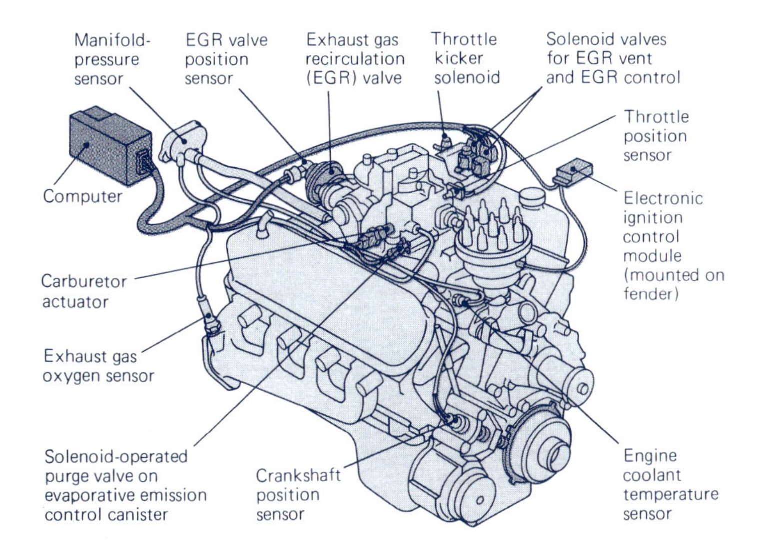

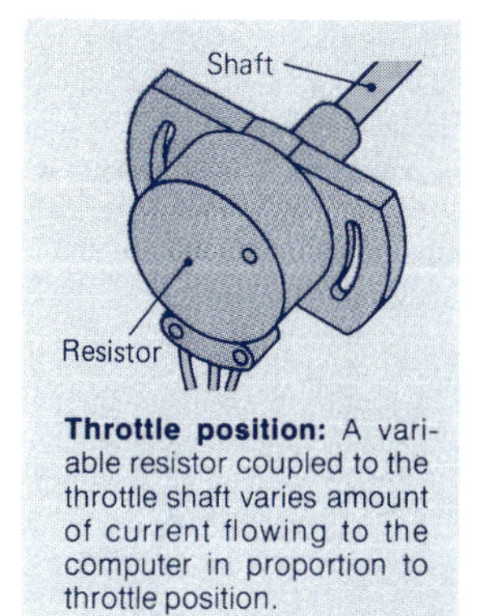

Throttle position: A variable resistor coupled to the throttle shaft varies amount of current flowing to the computer in proportion to throttle position.

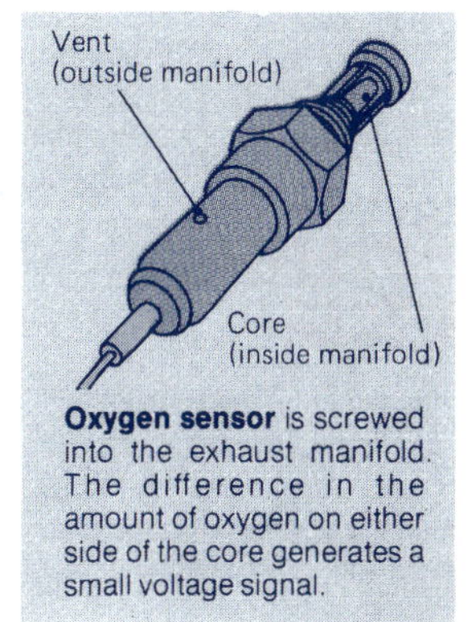

Oxygen sensor is screwed into the exhaust manifold. The difference in the amount of oxygen on either side of the core generates a small voltage signal.

4 Signals from the sensors

Look at the engine drawing
and listen to the interview on cassette.
Tick the components they mention.

5 Monitored and operated

Read the descriptions and fill in the gaps in the chart below.
Use words like **monitor**(ed), **control**(led), **operate**(d), **regulate**(d).

○ Air-fuel mixture	can be	....	by a motor on the carburettor. ○
○ Engine idle speed		....	by a diaphragm solenoid. ○
○ Exhaust gas recirculation		....	by a solenoid valve. ○
○ Crankshaft position		....	by a magnetic sensor. ○
○ Throttle position		....	by a variable resistor. ○
○ Amount of oxygen in the exhaust gas		....	by a voltage sensor. ○

Listen to the interview and tick the words and expressions they use.

Triple-Electric Ⓐ

TOKYO, JAPAN—So far, electric cars have run on either batteries, fuel-cell power or solar energy. Now, Sanyo has built a $400,000 2-seat concept car that taps all three power sources.

Amorphous-silicon photovoltaic cells, with efficiencies between 5% and 7%, coat the car's top surfaces.

The fuel cell is a small 250-watt powerplant weighing roughly 66 pounds. Three nickel-hydride cells, each weighing about 6 pounds, supply the hydrogen fuel.

The backup battery pack is made up of familiar nickel-cadmium units.

The car's top range on a sunny day is 94 miles, but only 30 miles without solar power.

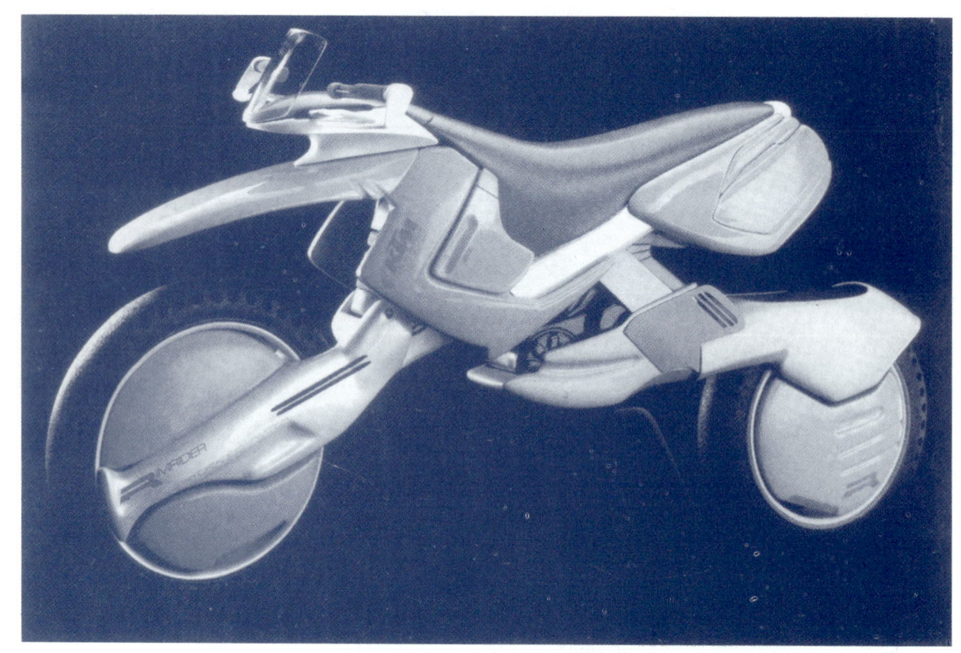

Stylish Cycle Ⓑ

The front wheel of the Rim Rider revolves around a stationary rim with no center hub, improving handling and reducing noise. The emissions system incorporates electronic synthesizers that reduce exhaust noise. Next World Design, 6566 Retton Rd., Reynoldsburg OH 43068.

Flywheel Trolley Ⓒ

When this electric trolley brakes, it spins a 400-pound flywheel that stores excess kinetic energy. When the trolley accelerates, the flywheel releases the energy so 25 percent less power is needed from the overhead cables. The flywheel is located next to the right front wheel. The 58-foot-long trolley is made by Neoplan. Gottlob Auwarter GmbH & Co., Box 810140, Vaihinger str 118-122, 7000 Stuttgart 80, Germany.

Sanyo's electric car draws power from photovoltaic cells, fuel cell and batteries.

1 Stylish cycle

Look at the photos and descriptions on the opposite page.
Which text describes which photo?
Write the appropriate letter next to each photo.

What is each of the products called?
Choose from the following names:
trolley (bus), sports car, dirt bike.

2 Flywheel trolley

Read the descriptions carefully and answer the following questions.

Which of the products . . .

. . . run on electricity? ....

. . . has a range of 94 miles a day? ....

. . . is about 17 metres long? ....

. . . is made in Japan? ....

. . . is made in Germany? ....

. . . is made in USA? ....

. . . has a front wheel revolving around a stationary rim? ....

. . . has a 200-kilo flywheel between the front wheels? ....

. . . has a synthesizer to reduce exhaust noise? ....

Compare your results with a partner.

3 Triple electric

Study the description of the electric sports car.
Which power sources does the car use?

○ gasoline (US)	○ solar energy
○ petrol (GB)	○ fuel cells
○ methanol	○ batteries
○ hydrogen	○ electricity

4 Rim Rider

Look at the photos on the opposite page and listen to the product presentation on cassette.
Which product is being presented?

....

Look at the following features and listen again.

- ○ Major parts made of metal alloys.
- ○ Engine block and frame made of carbon fibre and hi-tech plastics.
- ○ Chain-drive system.
- ○ Hydraulic rear-wheel drive system.
- ○ Chrome axles, spokes and hubs.
- ○ Front rim solidly fixed to swing arm.
- ○ Tires revolve around stationary rims.
- ○ Air-filled tires.
- ○ Solid tires.
- ○ Long muffler pipes.
- ○ Noise reduced by synthesizer.
- ○ ABS front brakes.
- ○ Hydraulic steering system.
- ○ Five-speed foot-operated shift gear.
- ○ Automatic transmission.
- ○ Air-cooled fuel-injected 2-stroke engine.

US: tire
GB: tyre

Tick the features **Rim Rider** will have.

5 Fixed to the front swing arm

Look at the parts in the list above.
Which ones can you see in the photo?
Show and explain them to a partner.

Here's the front rim.
It's solidly fixed to the ....
....

Which of the features listed above belong to traditional motorbikes?

Discuss the differences between **Rim Rider** and traditional bikes in a group of three.

Unit 37 Personal robots

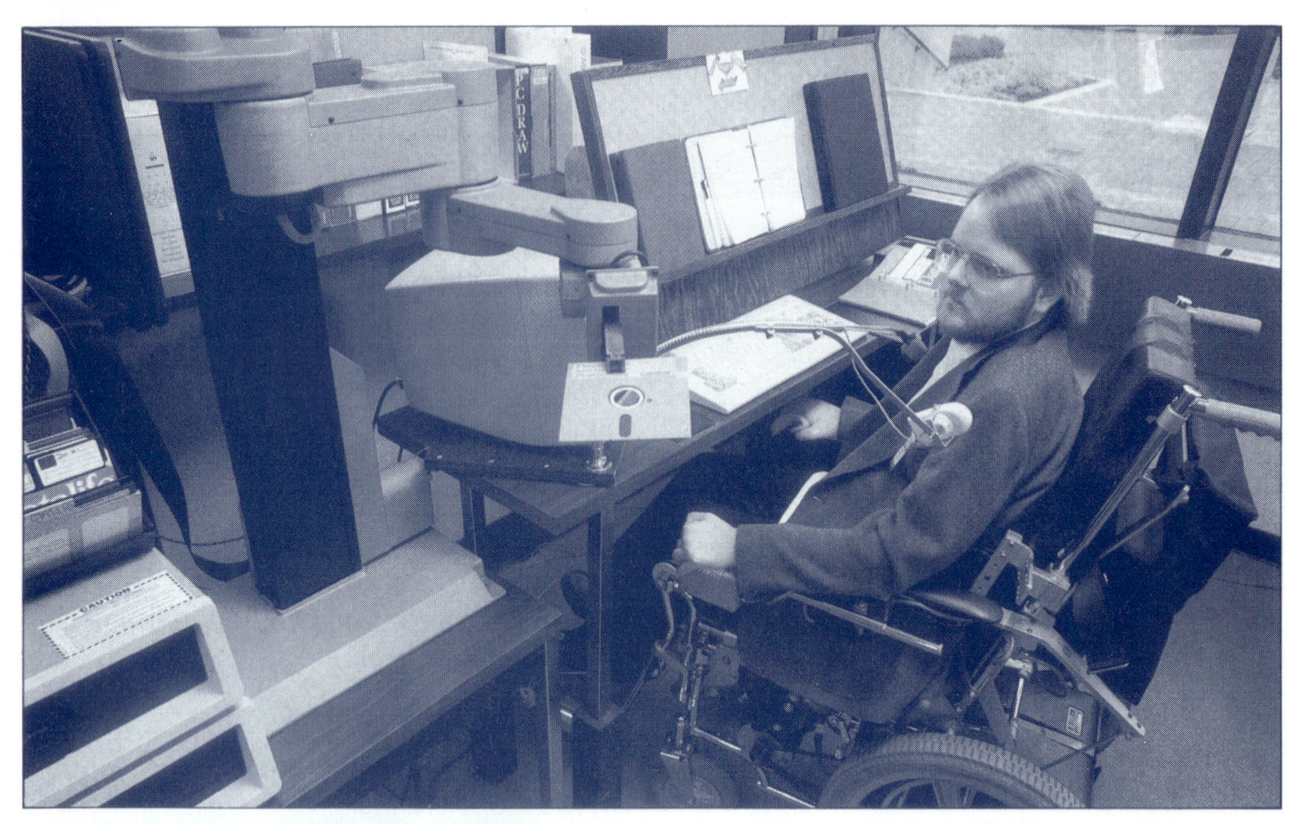

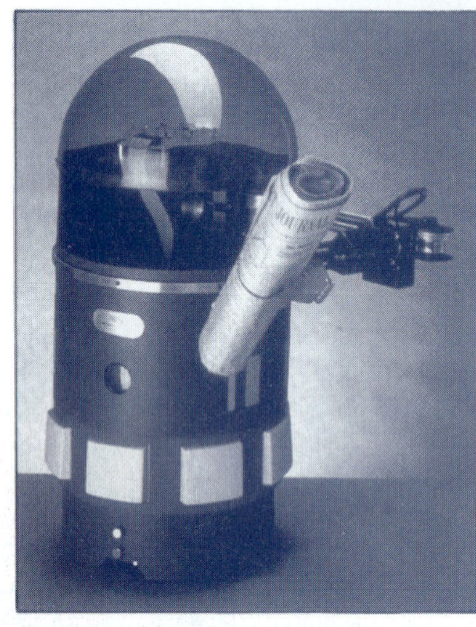

Meldog has been developed by the Japanese to help blind people. It has TV and other sensors to recognize the area ahead. Its computer compares the information from the sensors with the map stored in its memory and then sends a signal to the human owner. This machine has yet to prove itself more useful than a dog.

RB Robot's RB5X is designed specifically for use in the home. Around the base of the robot are square-shaped collision detectors. Once in contact with an object, these collision detectors tell the robot's on-board computer to change instructions and steer the robot clear of the obstacle.

The RTX robot built by Universal Machine Intelligence can be used in industry, in the laboratory and to help the disabled. It is being used in the United States to set up a workplace for people in wheelchairs. In this picture the RTX is bringing a floppy disc to the computer.

1 Robots at home?

Look at the photos on the opposite page and read the passages.
Which passage belongs to which photo? Write the letters A, B or C into the photo.

Look at the photos and study each passage.

The **MELDOG** has been developed
○ by a Japanese company.
○ as a robot "guide dog".
○ to recognize the area ahead.
○ to help blind people.

Discuss your results in a group of three.

The **RB5X** is designed
○ to be used at home.
○ to make contact.
○ to bring the mail.
○ to cook meals.

The **RTX** is used
○ to help disabled people.
○ to set up workplaces.
○ in industry.
○ in laboratories.

2 A useful application

Look at the photo on the opposite page and listen to the interview with Mr Devol who is an expert on robot applications.
Tick the photo they are talking about.

Which robot applications do they discuss? Take notes and compare them in groups of 3.

....

Listen again and connect the following statements.

Cleaning up or cooking seem ○
Robots would require ○
This robot is ○
It's ○
This is a workplace ○
That looks ○
There are wheelchairs ○

○ programmed with a map.
○ to be simple tasks.
○ millions of instructions.
○ for a disabled person.
○ fitted with ultrasonic sensors.
○ fitted with small robot arms.
○ like a useful application.

Compare your result with a partner.

3 An interesting development

Listen to this part of the interview and complete the text in the bubbles.
Work with a partner.

But how can .... help people?

Other than in ....?

Yes.

Well, here we have an interesting development for .... This robot is .... with a map and it's fitted with ....

Which applications for robots do you know?
What are the robots used for?
What else can you say about them?

Fiat factory in Italy uses unmanned trolleys to carry body shells around the factory. The trolleys follow a guide wire buried in the factory floor.

Staff fear shutdown at factory

by Jill Wells

STAFF at the trouble-hit STC factory in Brighton fear for their jobs after a disastrous six months for the company.

The telecommunications and computer group lost £8.7 million in the first half of this year and there is still no sign of trade picking up.

Now staff at the Hollingbury Estate fear their jobs are on the line and rumours of the plant closing are rife.

The Brighton operation has been run down dramatically over recent years.

In 1984 its 900-strong workforce was halved, while in April this year 360 manufacturing workers were given redundancy notices, leaving 130 people in the design and engineering division.

Of these, another 35 have been made redundant and are due to leave by November.

Mr Alf Verrall, secretary of the AUEW/TASS unions at the factory, said the latest dismal trading report had done nothing to improve morale at the plant.

He said: "The loss had been expected for some time but we are always nervous. We are in trepidation of losing our jobs."

He added: "The possibility of closing down is quite great, despite assurances from local management we will not. It is a very worrying time for everyone here."

But a spokesman for STC said plans for the Brighton plant would not be decided until the new chairman of the group, Lord Keith of Castleacre, had examined the firm's performance.

1 Staff fear shutdown

Read the article and answer the questions.

What is it about?

It's about
- ○ STC being closed.
- ○ trouble at STC.
- ○ redundancies.
- ○ financial losses at STC.
- ○ telecommunications and computers.

How many people worked at STC

before 1984? ....

After 1984? ....

Before April of this year? ....

How many work there now? ....

How many will be left in November? ....

What are the reasons given for closing STC
- ○ A new chairman.
- ○ STC has been run down.
- ○ Staff has been leaving.
- ○ A bad trading report.
- ○ Losses of £ 8.7 million.

How do the staff feel about it?
- ○ They fear for their jobs.
- ○ They are nervous.
- ○ They are glad to be out of work.
- ○ They are afraid of losing their jobs.
- ○ They are worried a lot.

Compare your results with a partner.

2 Redundant in six weeks' time

Listen to the interview with Ray Ede.
Tick all the statements which you think are correct.

What's going to happen at STC Brighton?
- ○ STC is part of a rationalization plan.
- ○ The company will be run down.
- ○ The factory is being closed.
- ○ The people will be redundant in six weeks' time.

Why?
- ○ Because they are not capable of producing any more.
- ○ Because of changes in technology.
- ○ Because of various other factors.

What do the following abbreviations stand for?

AEUW ○		○ Standard Telephone and Cables
STC ○		○ Technical and Administrative Services Section
TASS ○		○ Amalgamated Union of Engineering Workers

3 A lot of good people

Listen to the interview again. What can you say about Ray Ede?

How long has he been working for STC? ....

What does he like about STC?

○ The rationalization plan.
○ The amount of expertise.
○ The good people.
○ Their production capability.
○ Automated processes.
○ Modern technology.

How does Ray Ede feel about the closing of STC?

He is	angry ○		because	○ STC has a lot of good people.
	disappointed ○			○ he hates to see the expertise going somewhere else.
	indifferent ○			○ STC is at the peak of their production capability.
	philosophical ○			○ companies are closing all over the world.
	glad ○			○ it is a fact of life.

Compare your results with a partner.

4 Drawback calculated

Listen to Ray Ede once more and fill in the gaps.

Five years ago it used to take us ....
.... to produce one teleprinter.
It now takes us ....
because of ....
So .... it we are using the same number of people,
we have to sell nearly .... machines.
Or: if we sell the same .... machines,
we can do it with .... of the labour force.
And this is one of the ....
of ....

5 What about you?

How long have you been working for your company? ....

In which department? ....

What is being done there? ....

What do you like about the department? ....

Do you know any hi-tech companies? ....

Are there any rationalization plans? ....

How do you like your job?

Barry Brindle prides himself on working on the practical side of the testing business. As a youngster he had always enjoyed rebuilding and selling old bikes and even now, as a skilled tester, his real satisfaction comes from dealing directly with engine testing problems.
So his present job as a field tester for diesel engines gives him every opportunity to make use of his skills and his interests.

"Basically, my job is to plan and run series of tests on prototype machines to make sure that the engines are capable of the work the equipment demands. Our engines go into many types of vehicles and production machinery. When customers come to us they specify the operating conditions for the engine they need. So when the engine team has come up with a prototype, my job is to test it to make sure it satisfies both the customer's and our own specifications."

1 Barry's job

Read the passage on the left quickly.

What does Barry do?

....

Read the passage again and answer the following questions:

Where does Barry work?

....

How does he like his job?

....

What does he say about his company?

2 Testing prototypes

Listen to Barry.
What is he talking about?
Tick the right statements.

Barry
- ○ plans prototype test series.
- ○ designs prototype engines.
- ○ runs test series on prototypes.

The prototype should
- ○ be capable of the work the equipment demands.
- ○ satisfy the customer's specifications.
- ○ satisfy the company's specification.

The engines go into
- ○ tractors.
- ○ air planes.
- ○ forklift trucks.
- ○ combine harvesters.
- ○ production machinery.

Compare your results in your group.

SELF-ASSESSMENT QUIZ
Score yourself on these points.

I find ...	1 interesting	2 easy	3 boring	4 difficult	Friend's assessment
getting up in the morning	❑	❑	❑	❑	....
working with tools	❑	❑	❑	❑	....
reading drawings	❑	❑	❑	❑	....
technical lessons	❑	❑	❑	❑	....
metalwork	❑	❑	❑	❑	....
my colleagues	❑	❑	❑	❑	....
technical drawing	❑	❑	❑	❑	....
working on machines	❑	❑	❑	❑	....
assembling components	❑	❑	❑	❑	....
working in a group	❑	❑	❑	❑	....
working on my own	❑	❑	❑	❑	....
being flexible	❑	❑	❑	❑	....
working in design	❑	❑	❑	❑	....
looking for faults	❑	❑	❑	❑	....
repairing faulty equipment	❑	❑	❑	❑	....
testing components	❑	❑	❑	❑	....
wearing working clothes	❑	❑	❑	❑	....
being punctual	❑	❑	❑	❑	....
working on computers	❑	❑	❑	❑	....
wearing safety gear	❑	❑	❑	❑	....
installing equipment	❑	❑	❑	❑	....
being reliable	❑	❑	❑	❑	....
working in tight spaces	❑	❑	❑	❑	....

3 Self-assessment

Go through the points in the Self-Assessment Quiz and score yourself.

Then ask a good friend to assess you:

Take notes in the space provided.
Do you agree with his or her assessment?

4 What about you?

Answer the following questions and write a short report.

Do you like your school? ....

And your job? ....

How do you get on with your colleagues? ....

Did you attend a practical training course? ....

What was the most enjoyable part of your training? ....

....

Interview your neighbour and take notes. Then report to the class.

WANTED

BRITISH GAS

Invite applicants for the following vacancies:

DESIGN AND DEVELOPMENT ENGINEERS	C & G minimum. CAD training will be provided but previous exp. an advantage.
LABORATORY TECHNICIANS	College graduates preferred. Experience not necessary.
MACHINE OPERATORS	C & G, exp. preferable. Knowledge of CNC drilling and milling machines an advantage.
MECHANICS	Apprenticeship and further training offered.

An attractive salary and benefit package will be offered for each position.
Applications with CV should be sent to:

Fiona Grouse, British Gas Recruitment Office
152 Grosvenor Road, London SW1V 3JL, Telephone: 071 834 2000

Birmingham
TOOL AND DIE MAKERS
required for various tasks in our machine tool center. Age range 18–35. C & G min. Apply to C. Dugard Ltd, 75 Old Shoreham Road, Hove, Sussex BN3 7BE.

STEEL FABRICATORS
require
QUALITY CONTROLLER
Dynamic person required with salary to match
Box P615, Argus Brighton

South East
AUTOCAD DRAUGHTSPERSON
We require AUTOCAD persons for various disciplines for on-going and forthcoming long/short term contracts. Must have some experience with AUTOCAD Further training given. CV's please with details of experience to Pipco Ltd, 234 Hagley Road, Hasbury, Halesowen, West Midlands, B63 4QQ.

PLUMBER
required to work with heating. Site work. Attractive salary and benefit package. Phone Greg Mason, London
Tel. 081 579 7322.

MECHANICAL FITTER
required for temporary position. Must have some experience. Self-reliant. Immediate start. Phone Jill on Brighton 205281.

TALBOT
Seeks welders for new range of cars. Benefits, training given. Prospect of promotion.
Application forms are available from Personnel Department Talbot 14 Connaught Place London, W2 2ED
081-963 8089

ENGINEERING
INSTALLATION/SERVICE ENGINEER
Required for work in London & South east. You should have experience in intruder alarms. Knowledge of CCTV and access systems useful.
Please send your CV to Pat Felghan,
Rellance Electronics Limited, Suraty House, 86 Brigstock Rd, Thornton Heath CR7 7JA, 081 665 7473.

WADHAM KENNING, BRIGHTON
require an additional
SKILLED PAINT SPRAYER
The ability to use 2 pack products is essential. We offer good working conditions and all usual benefits associated with a large progressive company.
If you would like to join our friendly team then ring Mr M. Cooper, Service Manager on Brighton 821160.
WADHAM KENNING
NEWTON ROAD, HOVE

Hong Kong & London
TRAIN SYSTEMS DESIGN
require mechanical engineer to assist in developing the concept and system for Bangkok Metro. Minimum qualification C+G. Salary negotiable. Please quote Job No 48 (4), F. J. Selleck Assoc, (UK) Ltd, Int'l Recruitment Consultants, Prospect House, 17 North Hill, Colchester, CO 1 1DZ (Agy)

1 Work in England

Many young Europeans complete their English language studies in England.
Why not learn English while you are working abroad?

Look through the job advertisements. Tick the jobs that are offered and list the name of the firm, where it is (city) or any other information you find interesting.

Job offered	Company, place	Further information
○ Draughtsman/woman	....	....
○ Mechanic	....	....
○ Mechnical fitter	....	....
○ Service technician	....	....
○ Mechanical engineer	....	....
○ Welder	....	....
○ Design and development engineer	....	....
○ Paint sprayer	....	....
○ Tool and die maker	....	....
○ Machine operator	....	....
○ Quality controller	....	....
○ Laboratory technician	....	....

2 Job scene

Job scene is a telephone service provided by the Brighton Job Centre.

Listen to the recording and take notes in the following chart.

Job offered	Employer	Further information

Compare your results with a partner.

3 Working abroad

Which job do you find most attractive?

....

Why?

I'd like to work in London.

Good teamwork is important.

Quality control sounds interesting.

Tell your group about **your** reason(s).

Unit 41 And your education?

1 A new job

Many young people leave their home countries to take up jobs abroad. Working in a foreign country can be an attractive option for a number of reasons.
Before you take up a new job you should consider some things. Number the steps on the right. They will lead you from where you are to where you want to be.

- ◯ Job interview
- ◯ An appointment
- ◯ Letter of application
- ◯ Curriculum vitae
- ◯ Job advertisements
- ◯ Calling or visiting companies

Curriculum Vitae

Catherine Bainbridge
1, Hollin Wood Close
Moorhead
Shipley, West Yorkshire
0274 595978

born 9 October 1976, in London

Parents	John and Beryl (Bazen)	
Education	Primary School	1981 - 1987
	"Beckfoot" Comprehensive School Wagon Lane, Bingley	1987 - 1992
Qualifications	GCSE Mathematics, Physics, Chemistry, Sports, English Language, Design and Technology.	1992
Special Training	Evening classes in "Technology Today" and "Environmental Design".	1992
Work experience	Weekend work at a local garage. General duties.	1989
	Helped design stage set for University production at theatre.	1990
	Trainee at Flymo lawnmower manufacturers research department, Darlington	1992 - 1993
	Fitter, specialising in machine tool cell integration, KTM, Brighton	1993 to present
References	Dr. Paul Abbey (employer) Flymo research department Darlington, England Telephone 0942 324774	
	Peter Albee KTM Ltd, Garden Avenue Brighton, England Telephone 507255	
Interests	Jiu-jitsu, Cycling, Model-making	

Shipley, May 1995 *Catherine Bainbridge*

2 Curriculum Vitae

Curriculum Vitae (CV) is Latin for "course of life". It is a personal history chart that helps to structure information about yourself.

It can be used repeatedly for many job applications. So, it should be easy for employers to read.

Read the CV and tell a partner what it says about Catherine Bainbridge.

"I warned you not to get it done before they've accepted you."

3 Schools in England

Look at the diagram and listen to Howard.
Fill in the ages.

BTEC . . .	**B**usiness and **T**echnician **E**ducation **C**ouncil
BTEC First Certificate	One year full-time or two year part-time course
BTEC National Certificate	Two years of full-time study or three years part-time
C&G . . .	City and Guilds Diploma
"A"-level	Advanced Level, exams in three or four subjects of your choice.
GCSE . . .	**G**eneral **C**ertificate of **S**econdary **E**ducation: Compulsary examinations in English language and literature, mathematics, science, and a foreign language.

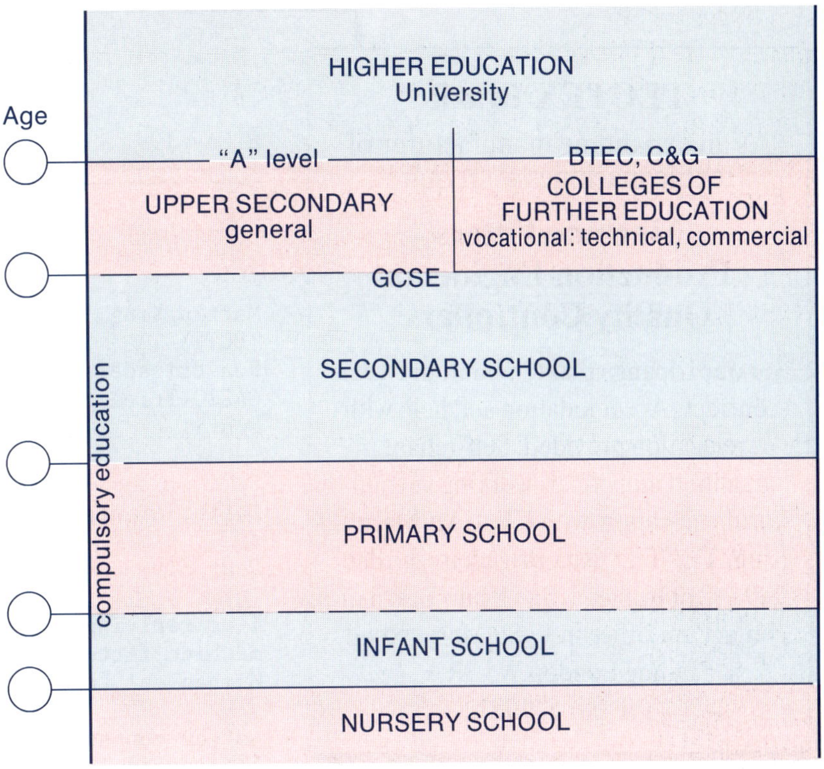

Age

HIGHER EDUCATION
University

"A" level — BTEC, C&G

UPPER SECONDARY general | COLLEGES OF FURTHER EDUCATION vocational: technical, commercial

GCSE

compulsory education

SECONDARY SCHOOL

PRIMARY SCHOOL

INFANT SCHOOL

NURSERY SCHOOL

4 Catherine Bainbridge

Read Catherine's CV again: Tell your partner the answers to the following questions.

When and where was Catherine born?
Who are her parents?
Which schools did she attend?
When?

What are her qualifications?
Has she got any further training?
Where did she work?
What does she do now?

5 Schools in Germany

How do British schools compare to German schools?

Think of the schools you went to and find the English words.

Hochschule ○	○ University
Berufsschule ○	○ College of Further Education
Fachschule ○	○ Technical school
Realschule ○	○ Vocational school
Fachoberschule ○	○ Grammar school
Gymnasium ○	○ Commercial school
Gesamtschule ○	○ Primary school
Hauptschule ○	○ Secondary school GCSE
Grundschule ○	○ Secondary school "A" Level

6 Your CV

Tell a partner in English which schools you attended. Consider the following examples.

From .... to .... Then After that	I went to I attended	primary school secondary school	in ....
And now	I go to	the technical college the vocational school for	in ....

Write your own curriculum vitae on a separate sheet of paper.
First, write a draft, so that you can see how much room you need.
Choose a neat layout and print the final version.

Yours sincerely

TECTEX Brazil

As an expanding manufacturer of textile machines we require

Machine Fitters
Production Engineers
Quality Controllers

for our Iripanga plant, minimum 2 year contract. Accomodation and help with relocation provided. Self-reliant, qualified individuals working to train local team apply with CV to Mr Martin King, TECTEX Rua Brigadeiro Jordao, 149 04210 Iripagna, Sao Paulo Brazil or contact our office in 17 St. Marys Road, Ealing London W5 5RA
Tel. 081 579 7222

Catherine Bainbridge
1, Hollin Wood Close
Shipley, West Yorkshire

8 May 199.

Martin King
TECTEX
Rua Brigadeiro Jordao, 149
04210 Iripanga, Sao Paulo
Brazil

Letter of application

Dear Sir,

I am replying to your advertisement for a machine fitter in the March issue of Mechanical Engineering Today magazine.

As you can see from my curriculum vitae (enclosed) I am working as a machine fitter for KTM.
I have some experience in the field of Machine Tool Fitting and I would like to gain some international experience.

I hope that you can consider me for the post at your company in Brazil.

Yours sincerely

Catherine Bainbridge

Catherine Bainbridge

1 Letter of application

Read the letter of application and find answers to the following questions:

Who is writing the letter?

Who is it addressed to?

What's the reference?

What does the sender want?

2 Dear Sir, Madam

Read the guidelines for applying by letter and check Catherine's letter. Correct it if you find a mistake.

WHAT IT SHOULD SAY

- If you know the name of the person you're writing to, put it in.
- If not, write 'Dear Sir' or 'Madam'.
- Say what job you're applying for and where you saw it advertised.
- Say why you're interested in the job and if you have any useful experience.
- Say what other skills you have or any exams you've passed.
- Say that you've enclosed more information on your CV.
- Put in something like, 'looking forward to hearing from you.'
- If you put someone's name at the top, you should sign the letter 'Yours sincerely',
- If you put 'Dear Sir' or 'Madam', write 'Yours faithfully.'

Always try to get someone else to check your letters

WHAT IT SHOULD LOOK LIKE

Address – put yours in the top right-hand corner and theirs on the other side, a little lower down.

Date – the date should go about an inch under your address.

Make it neat – no crossings out or smudges, if possible.

3 Motorcycle mechanic

Look at the advertisement.
How would you reply to it?
Read the parts of the letter and sort them out.
Number the parts from 1 to 10.

C. DUGARD LIMITED
MACHINE TOOLS
require
production technician BTEC
machine fitters and **welders C&G**
for machine tool production

75 Old Shoreham Road, Hove, Sussex BN3 7BE
Telephone Mr Clisby 0273-728851
Telex 877423 Fax 0273-203835

13 May 19. .

Letter of application

As you can see from my curriculum vitae (enclosed) I am preparing for the City and Guilds diploma in machine fitting.

I am German and I would like to work in England to gain international experience.

C. DUGARD LIMITED
Machine Tools
75 Old Shoreham Road
Hove, Sussex BN3 7BE
Great Britain

I very much hope that you can consider me for one of the posts at your company.

I am replying to your advertisement for machine fitters in the April issue of 'Popular Mechanics' magazine.

Dear Mr Clisby,

Martin Marek
Sophienweg 24
90559 Burgthann
Germany 09183-17877

Yours sincerely

Martin Marek

4 International experience

Remember the job you liked best in Unit 40?
Write your own letter of application using passages from the letter here.
Good luck!

5 Get organised

Listen to this bit of advice from Job Scene.
What do they recommend?

- ○ Get organised.
- ○ Make a list of interesting firms.
- ○ Contact the firms on the phone.
- ○ Write down the dates of your letters of application.
- ○ Record the dates of replies.
- ○ Ask a parent to apply for you.
- ○ Follow up letters in case of no reply.
- ○ Fill in application forms.
- ○ If you don't get an answer just drop in for a visit.
- ○ Prepare for interviews.

Can you make it?

1 Can you make it at nine?

Listen to the telephone conversation. When and where will the interview take place?

....

2 I'd be delighted

Listen again and fill in the gaps.

I'm calling about ....
....
We're ....
Can you ....

I'd be ....

You understand ....
.... applicants.

Yes, ...
So, that's ....
at .... at ....?

Right, at St. Mary's Road.

I've got ....

We'll see you at ....
on ....

Essex
LIFT ENGINEER

To service/maintain lifts, escalators and associated control gear. Must have previous experience. Ref J4178. For more details of this and other engineering positions telephone 0206 549845 or send c.v. to Compass Employment, The Moel Centre, 6 Grange Way, Whitehall Ind Estate, Colchester, Essex, CO2 8HF. (Agy).

South East
MILLWRIGHTS, MACHINE TOOL FITTERS AND MECHANICAL FITTERS

We require Millwrights, Machine Tool Fitters and Mechanical Fitters for forthcoming Long/Short Term Contracts. Apprentice Trained, C & G Qualifications. C. V's please to PIPCO Limited, 234 Hagley Road, Hasbury, Halesowen, West Midlands, B63 4QQ

3 Lift engineer

When telephoning employers always cover the following five points. The order can be varied.

Why are you looking for work?

What type of work are you looking for?

What skills do you have to offer?

Why are you calling that firm?

What do you want from the firm?

Look at the advertisements on the left or look up the company you chose from Unit 40.
Prepare a telephone conversation with your partner.
He or she can be your future employer.

4 Dear Mr Marek

Read the letter to Martin.
What do you think of it?
Would you contact Ms White?
What would you say
on the telephone?

C. DUGARD LIMITED
MACHINE TOOLS
75 OLD SHOREHAM ROAD, HOVE, SUSSEX BN3 7BE
Telephone: Brighton (0273) 728581 - 732286
Telex: 877423 Fax: (0273) 203835

Martin Marek
Sophienweg 24
D- 90559 Burgthann
Germany

12 May ....

Re: **Your application**

Dear Mr Marek,

We are very pleased with your application and with your broadbased training in metalwork and machine fitting.
Our chief engineer, Ms White, is on a business trip in Germany and could arrange to meet you for an interview. Please contact her on Monday, 1 June, at the Hotel Europa in Nuremberg, telephone 515940.

I wish you all the best for your interview.

Yours sincerely,

Peter Clisby

Peter Clisby
Production Manager
C. DUGARD LIMITED

Juni/June/Juin

1	Fr/Fr/Ve	
2	Sa/Sa/Sa	
3	So/Su/Di	
4	Mo/Mo/Lu	
5	Di/Tu/Ma	
6	Mi/We/Me	
7	Do/Th/Je	
8	Fr/Fr/Ve	
9	Sa/Sa/Sa	
10	So/Su/Di	
11	Mo/Mo/Lu	
12	Di/Tu/Ma	
13	Mi/We/Me	
14	Do/Th/Je	
15	Fr/Fr/Ve	
16	Sa/Sa/Sa	
17	So/Su/Di	
18	Mo/Mo/Lu	
19	Di/Tu/Ma	
20	Mi/We/Me	
21	Do/Th/Je	
22	Fr/Fr/Ve	
23	Sa/Sa/Sa	

5 When can I come?

Now you can make an appointment for an interview.
Find a partner who wants to be the manager.
Call him or her and write the details
of the appointment into your diary.
Present the scene to the others in your class.

Confidential

Application for Engagement

CONFIDENTIAL

Position applied for	Source of introduction
Machine fitter	Popular Mechanics magazine

PERSONAL DETAILS

Surname Mr/Mrs/Miss			First names	
Marek			Martin	
Permanent address			Present address (if different)	
Sophienweg 24 D-90559 Burgthann Germany 09183-17877				
Age	Date of birth	Place of birth	Nationality	Nationality at birth
18	20. Mar. 1975	Nuremberg	German	
Health. Please give details of any serious (or recurring) illness or injury you have had.				
Are you a registered disabled person? No		If yes please state Reg. No.	Nature of disability	

NEXT OF KIN

Name	Address
Relationship	Tel. No.

EDUCATION

Name & type of schools	Dates	Examinations taken. Please give results and details of any certificates, awards etc.
		....

INTERESTS

Please give details of interests, sports and hobbies, including memberships committees and clubs etc.

EMPLOYMENT HISTORY (Start with present or last employer and include any service in H. M. Forces)

Dates	Name, address and business of Employer	Positions held Give brief description of duties	Wage or Salary	Reason for leaving
		....		

If you have applied to this company before, give dates and position applied for. .... Date available to take up new employment	I confirm that the information supplied in this application is correct and I understand that this employment is subject to the receipt of satisfactory references. I also confirm that I have not withheld any information, the declaration of which might prejudice my acceptability for employment within the Company. Signature .... Date ....

1 Some international experience

Look at the Application Form on the opposite page and listen to the job interview. Which parts of the form are they talking about? Listen again and complete the form wherever you can.

2 Bright and alert

Listen again.
If you had a firm would you hire Martin Marek?

Which of the following do you think apply to Martin?

- ○ bright and alert
- ○ fit and strong
- ○ serious
- ○ punctual
- ○ relaxed
- ○ friendly
- ○ good looking
- ○ shy
- ○ neatly dressed
- ○ laughs easily
- ○ competent
- ○ modest

3 Quite the contrary

Read the questions and answers in the bubbles.
Do you remember who said what?
Mark the bubbles Ⓜ for Martin and Ⓦ for Ms White.
Work with a partner.

You aren't married, ...?

How does that work in Germany?

For how long?

You have to take an exam.

Why? Is anything wrong?

Why do you want to come to England and work for us?

Yes, something like the City and Guilds examination in England.

No, not at all. Quite the contrary.

I'd like to get some international experience.

But your English is very good.

And I'd like to work on my English.

Which of the questions relate to an answer?
Read them in sequence.
Your partner can be Mr Cook.

"Yes, you meet all my requirements . . . now for the firm's."

4 Job interviews

Find a partner who wants to interview you.
Agree on the details of a job and some questions he or she should ask you.
Then sit for an interview.
Your partner should fill in an application form.

C. DUGARD LIMITED

MACHINE TOOLS

75 OLD SHOREHAM ROAD, HOVE, SUSSEX BN3 7BE
Telephone: Brighton (0273) 728581 - 732286
Telex: 877423 Fax: (0273) 203835

CONTRACT OF EMPLOYMENT

For:	Martin Marek
Position:	Machine fitter
Department:	Manufacturing and maintenance
Reporting to:	Senior technician
Wages:	125 Pound Sterling per week
Working hours:	38 hours per week Monday to Thursday 8 - 12 am and 1 - 5 pm Friday 8 am - 2 pm
Job description:	Sub-assembly and final assembly of machine tools; test runs; installation and maintenance of machinery on customer's location; driving company cars with care.
Holidays:	20 days per annum, six months after commence of work
Social benefits:	sickleave 10 days per annum; social security and retirement benefits paid; in-company training free of charge.
Notice:	six weeks, to be given by either party.

Date signed: ....

Peter Clisby   Martin Marek

....

Peter Clisby
C. DUGARD LIMITED

Martin Marek

1 Machine fitter

Martin Marek has got a new job. Look at his Contract of Employment on the opposite page and ask a partner the following questions. You need not write anything.

Which firm does Martin now work for?

What's his job title?

Which department does he work in?

Who is he responsible to?

How much does he earn?

How many hours per week does he work?

When does he stop work on Wednesdays?

And on Fridays?

What are his duties and responsibilities?

How many holidays per year does he get?

And when at the earliest?

What sort of social benefits does he get?

How much notice does he need to give?

2 My job description

Now think of your own job. Answer the questions yourself and take notes.

My job:

I work for ....

....

....

....

....

....

....

....

....

....

....

....

....

3 A contract of employment

Interview a partner about his or her job. Refer to the questions about Martin Marek and take notes on a separate piece of paper. Then write a **contract of employment** for your partner.

Compare your job with your partner's.

He/She earns **more than** I do.
He/She works **fewer** hours **than** I do.
I have **longer** holidays **than** he/she has.
His/Her boss is **taller than** mine.
He/She works **the same** hours as I do.
His/Her company car is **as** old **as** my company car.

4 The Jack of Hearts

The Jack of Hearts is the lucky card. And luck is what you need for a job interview, among other things.

Listen to these job interviews.

Cassette script

Unit 1 **1 Meet our speakers**

Melanie: Hi! I'm Melanie Dalton.
I'll tell you my real name
and a little bit about myself,
because you are going to hear my voice on this cassette quite often.
I'm from America, but I live in Darmstadt now.
I'm thirty-four year old and
I teach English at the Volkshochschule here.
On this cassette I'll also be working with Tom.
But he'll tell you about himself.

Tom: Hello! That's right: I'm Tom D'Agostino.
I'll spell my last name for you,
because people always get it wrong:
That's capital D-apostrophe-
capital A-G-O-S-T-I-N-O. D'Agostino.
Did you get it?
Okay. And I'm American too,
from Stamford, Connecticut.
Yeah, that's another difficult word, isn't it?
That's C-O-double N-E-C-T-I-C-U-T.
Got that?
Connecticut is one of the states in the USA of A: Connecticut.
Can you say that? Can you say that?

Marijana: Connecticut.

Tom: That's good!
But now over to Marijana ...

Marijana: Hi! My name is Marijana Dworski,
and I'm British, from Wales.
Dworski is a Polish name.
And my first name is Yugoslav: Marijana.
The spelling is a bit tricky, too:
M-A-R-I-J-A-N-A. All right?
I came to Cologne two years ago,
and I each here.
I'm twenty-nine and I'm single.
You'll be gettint to know me
as we go along.
And here is one more colleague on the cassette: Howard.

Howard: Yes, my name is Howard Nightingall.
And I'm from East Greenstead,
a town south of London.
How do you do?
Now to our work here: We all hope
that our cassette will help you
to learn English.
But you should have some fun as well ...

Dennis: Also, Sie sollen auch Spaß
an der Kassette haben.
Sie sehen, unser Deutsch ist auch nicht so gut,
obwohl wir schon so lange
in Deutschland leben.

Amy: Also, wenn Sie manchmal nicht alles
verstehen, was wir sagen,
hören Sie sich das Gespräch einfach öfter an.
Sie werden von Mal zu Mal mehr verstehen.
Wichtig ist uns, daß Sie selbst das Wesentliche heraushören.
Auch Ausländer wiederholen gerne,
wenn sie etwas nicht gleich verstehen.
Nicht wahr?

All: Yeah ... sure ... right.

Now listen again and look at the photo in your book.
Who do you think is who? Good luck!

Unit 1 **3 Meet my partner**

Look at the paragraph in your book and
listen to Howard: He'll introduce Marijana again.

Howard: This is Marijana Dworski.
M-A-R-I-J-A-N-A Dworski. Dworski?
Aha, that's D-W-O-R-S-K-I.

Marijana: But you can call me Marijana.

Howard: Oh, that's nice.
She's 29 years old.
She's an English teacher in Cologne.
She comes from Wales, but she's English.
And she's single.

Did you get that?
Listen again and fill in the missing words.

Unit 2 **3 Starting a Career**

Look at the photos in your book and listen to the interview.

Interviewer: I'm talking to Tina Brown. Hi, Tina.

Tina: Hi.

Interviewer: I'll be asking you about your job
and the company you work for.

Tina: Yes, that's okay.

Interviewer: So. Er ... you are an apprentice here?

Tina: Yes, at General Motors.

Interviewer: When did you start?

Tina: Four weeks ago.

Interviewer: And how long does your training take?

Tina: Three and a half years.

Interviewer: Aha. And what will you be then?

Tina: Ohh ... a mechanic, I hope.

Interviewer: Mhm. And are you trained exclusively at GM?

Tina: No. I'm in the shop for 4 days a week
and one day at the vocational school.

Interviewer: They have a one day course for motor mechanics there?

Tina: Yes: one day per week, ten months a year.

Interviewer: Which day is that?

Tina: Friday.

Interviewer: And they give you the time off at work??

Tina: Yes.

Interviewer: Ah. Er ... what do you learn at school?

Tina: Oh, a log of things: maths, technology,
metallurgy, safety procedures,
technical drawing, print reading ...

Interviewer: Any languages?

Tina: Yes, English.

Interviewer: And what do you learn at work?

Tina: All the skills I need for my job,
such as turning, drilling, milling, repairing cars ...

Interviewer: Okay Tina, thanks for the interview.

Which one of the people in the photos is being interviewed?
Listen again and answer the questions in your book.

Unit 3 **1 Sweets after all?**

Interviewer: Martin, your name is Martin, isn't it?
Martin: Yes.
Interviewer: Where do you work?
Martin: Which company you mean?
Interviewer: Right.
Martin: I work for Haas.
Interviewer: The food company? Sweets and baking powder?
Martin: Well, no. We manufacture machines for producing wafers.
Interviewer: What do you mean by wafers? What sort of wafers?
Martin: All kinds of wafers: large ones, small ones, cream-filled wafers, ice-cream cones and cups, wafer sticks and rolls ...
Interviewer: Ah, I see. So, it is sweets after all?
Martin: Well ... not really. It's the machines and production units for making wafers ...
Interviewer: And not the wafers themselves?
Martin: Right.
Interviewer: Mhm. So, how are the wafers made?
Martin: Well, for example, to make ice-cream cones, batter is filled into the top of small hand-operated machines and out come the ice-cream cones. But we have also got semi-automatic units and everything up to fully-automated, computer-controlled wafer production lines.
Interviewer: Wow, sounds impressive. I'd like to get some more information. Have you any leaflets or a brochure?
Martin: Certainly. We've got everything in English.
Interviewer: Thank you very much, Martin.

Where does Martin work?
What do they produce?
Listen again and answer the questions in your book.

Unit 4 **4 Plant layout**

Peter: Dear Martin,
Take the plant layout I sent you
and look at it.
Ready?
Now: From the main road, that is Old Shoreham Road, you come into what we call the number one building.
There we've got the reception area and a showroom and at the back there are all the offices for the managers, design and development engineers, draughtsmen and women and the office staff.
This building is connected to the number two building where we get the raw materials.
Next to that there is the machining centre which connects to building number three.
In building number three there are a spraypainting and washing cell,
the tool management section and the component assembly area.
All the way in the back of building number three there is a computer room for CNC components and panels.
Now, once the components are assembled they come back to the final assembly area in building number two.
Before the machines are loaded and shipped to the customers they are tested in the inspection cell at the end of the building number two.
Well, there you are.
I hope my explanation was not too complicated.
Could you follow it?
Maybe you'd like to send me a cassette as well.
I'd like to hear your voice.

Unit 5 **2 BDR time**

Listen to the speakers on the BDR morning show.
What are they talking about?

It's twenty-three minutes before eight o'clock on Blue Danube.
This is Earth, Wind and Fire ...
Coming up to eight minutes ...
or rather ten minutes before eight on Blue Danube.
It's time for our phrase of the day ...
Four minutes before the news on BDR.
At eight o'clock this morning.
Here we are: Thursday, the sixteenth of July.
At this day in 1439 kissing was made illegal in England in an attempt to stop the spread of diseases. All back in 1439 ...
But now the time: It's eight o'clock.
(Murray Hall): The next ORF news in English on the first programme is tomorrow morning at five past eight. For listeners to Blue Danube Radio there'll be more today at twelve noon.

Unit 6 **1 When will I have to leave?**

Marijana: Excuse me. Do you speak English?
Travel agent: Yes. How can I help you?
Marijana: I have to be in London on the 9th of July. What possibilities do I have?
Travel agent: At what time would you like to be in London?
Marijana: My course starts at nine.
Travel agent: Am?
Marijana: Yes.
Travel agent: Let's see ... Here: There's one that'll get you to London Victoria at six thirty in the morning.
Marijana: When will I have to leave from Munich?
Travel agent: Munich ... departure at twelve forty-four on the eighth of July.
Marijana: Will I have to change?
Travel agent: Yes, in Cologne. You arrive there at seventeen fifty-nine ...
Marijana: One minute to six.
Travel agent: Yes. And tou take the eighteen seventeen to Oostende where it connects to the Jetfoil.
Marijana: And I'll arrive in London ...?
Travel agent: Er ... at six thirty, on the 9th of July.
Marijana: Mhm. Er ... can I book a couchette?
Travel agent: Well, you can book at the ticket window, over there.
Marijana: Thanks very much.
Travel agent: You're welcome.
Marijana: Bye.

So, where's Marijana going? Where does she leave from?
Which train will she take?
Listen again and answer the questions in your book.

Cassette script

Unit 7	**1 The Landing Card**
Jetfoil announcement	Ladies and gentlemen, we will be landing in Dover at the Hoverport at 14.25 local time. Please, don't forget to set back your watches to British daylight saving time. The weather is fine, temperature around 72 degrees Fahrenheit, that's 18 degrees centigrade. We thank you for travelling Hoverspeed and wish you a pleasant stay in England.
Stewardess:	Have you filled in your Landing Card, sir?
Tom:	Do I have to?
Stewardess:	Yes sir, if you are from a country outside the European Union ...
Tom:	But my students don't need a Landing Card.
Stewardess:	Because they're Germany, and ...
Both:	Germany's a member of the EU (laugh).
Tom:	Look, I've got another problem.
Stewardess:	Oh, I'm sorry. I didn't see your arm.
Tom:	Here's the card.
Stewardess:	Can I fill it in four you?
Tom:	Yes, thank you very much. My name's D'Agostino. Here's my business card.
Stewardess:	D'Agostino. And your first name is Thomas?
Tom:	Right.
Stewardess:	And what does A stand for?
Tom:	A ist for Anthony.
Stewardess:	All right. And M for male.
Tom:	Thanks.
Stewardess:	When were you born?
Tom:	On the 2nd of August 1950 in Stamford, Connecticut.
Stewardess:	Stamford, with an M?
Tom:	Right.
Stewardess:	Connecticut. And you're American?
Tom:	Yes, and I'm a teacher.
Stewardess:	Teacher. And where in England are you going to stay?
Tom:	At a hotel in Brighton, the Madeira.
Stewardess:	Is that M-A-D-E-I-R-A?
Tom:	Yes, that's it. 19 Marine Parade, Brighton Sussex.
Stewardess:	Brighton Sussex.
Tom:	Right.
Stewardess:	Can you try and sign it with your left hand?
Tom:	Yes, I'll try. Thank you very much.
Stewardess:	That's okay.

Did you find out what happened?
Listen again and fill in the landing card for Tom.

Unit 7	**3 Can I see your passport, please?**
Amy:	Well ... I think we have to go our separate ways now.
Students:	Why?! What do you mean?
Buky:	Amy's American. She can't follow the sign for EU nationals.
Male student:	And we are European Union.
Another student:	We'll see you after passport control?
Amy:	Yes. Go along now.
Immigration officer:	Can I see your passport, please?
Amy:	Just a moment.
Immigration officer:	And your landing card?
Amy:	Here you are.
Immigration officer:	How long do you intend to stay in England, Ms Lindner?
Amy:	Three weeks.
Immigration officer:	Is this on business or holiday?
Amy:	I'm with a group of students on an English course.
Immigration officer:	So, your business is teaching?
Amy:	Well, I'm their teacher at home.
Immigration officer:	Your home is Munich?
Amy:	Right.
Immigration officer:	But you're American?
Amy:	Right.
Immigration officer:	Have a pleasant stay in England!
Amy:	Thank you.
Customs officer:	Is this your bag, madam?
Amy:	Yes.
Customs officer:	Would you open it for me, please?
Amy:	All right, but it's only clothes and personal belongings.
Customs officer:	And what's in this bottle?
Amy:	It's German wine, a present for a friend of mine.
Customs officer:	Mhm. All right, madam. Welcome to Britain.

Unit 8	**2 Confirming a reservation**
Tom:	Let's see. That's 0-2-7-3-6-9-8-3-3-1
Receptionist:	Madeira Hotel. Good evening.
Tom:	Good evening, this is Tom D'Agostino speaking.
Receptionist:	Yes sir, can I help you?
Tom:	Yeah, I just arrived at Dover and I want to make sure that you got my reservation.
Receptionist:	What did you say your name was?
Tom:	D'Agostino, Tom D'Agostino.
Receptionist:	Ah yes, Mr D'Agostino. A single room with shower, wasn't it?
Tom:	That's right. I'll be in Brighton by nine. Can you keep the room for me?
Receptionist:	Yes, certainly. We'll see you around nine then.

Where do you think Tom is now? Who is he calling?
Listen again and answer the questions in your book.

Unit 8 **3 Here's your key**

Tom has arrived at the Madeira Hotel.
Listen first to what's happening there.
Ready?

Receptionist: ... Here's your key, sir.
Room number twelve.
Tom: Thanks.
Receptionist: Would you like to register, please?
Tom: Er ...
Receptionist: Oh sorry.
Let me do it for you.
Let's see.
I have your name from your letter.
Mr D'Agostino, T.
And your home address, Mr D'Agostino?
Tom: Aachener Straße 51.
Shall I spell it for you?
Receptionist: Oh yes, please.
Tom: That's double A-C-H-E-N-E-R Straße fifty-one.
Receptionist: ... fifty-one.
Tom: Right. And that's Cologne.
Five, oh one six two Cologne.
Receptionist: 50162 Cologne, Germany.
How long will you be staying,
Mr D'Agostino?
Tom: Three or four nights.
Receptionist: Room number 12 is on the second floor.
Just follow the porter.
Tom: All right. Thank you.

So, Tom got to his hotel all right.
Could you fill in the guest register for Tom?
Start with today's date, then listen again.

Unit 9 **2 What sort of food do you like?**

Tom: Ah well ... Where should we go?
What do you think?
You're from here.
You know your way around here.
Marijana: Yeah, but we're not from Brighton.
John: I've got the Evening Argus here.
Let's have a look:
What sort of food do you like anyway?
Marijana: Erm ... I like Continental food.
Are there any French or Italian
restaurants here?
John: Yes. There's the French Connection,
Romano's, which is Italian,
and Christy's ...
Tom: Oh, we have enough Italian restaurants.
I feel like something more exotic.
John: What do you think of a Mongolian Barbecue
at Genghis Khan's?
Tom: Oh!
Marijana: Sounds expensive.
John: Or an Indian meal at the Tea Planters?
Tom: No, I don't think I can handle
a curry dish tonight.
Marijana: No.
John: You two are hard to please, aren't you?
Tom: Hey, I have an idea.
Have you heard of Browns?
John: Let me see.
They don't seem to advertise in the Argus.
Tom: Well, it's an interesting place.
They do American and Continental dishes,
some English food as well ...
Marijana: Yeah, why didn't you tell us about that before?
Tom: Oh, I didn't think of it.
John: Well, let's go there.

Now listen again and look at the Evening Argus Restaurant Revue.
Which restaurants do our friends talk about?

Unit 9 **4 Are you ready to order?**

Tom: Er ... you know what you're having, Melanie?
Melanie: Yes, I think I'll have ...
Waitress: 'you ready to order?
Melanie: Just about.
I'll have a Mrs Browns Vegetarian Salad,
please.
Waitress: Er ... Which dressing?
Melanie: Sorry?
Waitress: Well, you can have Blue Cheese, Thousand
Island, French dressing, Mayonnaise or Garlic
dressing.
Melanie: Ah yes. French dressing, please.
John: And I'll have the Fisherman's Pie,
with the vegetables, please.
Waitress: Right. And you, sir?
Tom: I'll try the Leg of Lamb.
With a baked potato.
Waitress: And which dressing on your side salad?
Tom: Thousand Island.
Waitress: Right.
Anything more to drink?
John, Melanie: No, thanks.

Listen again and look at the menu in your book.
Can you find the dishes our friends order?
Hm, makes me hungry.

Unit 10 **1 Here's your bill**

Waitress: Did you enjoy your meal?
John: Oh yes, thank you, it was delicious.
Melanie: Yes, it was very good.
Tom: Can we have the bill, please?
Waitress: Yes. Certainly.
Melanie: But what about desserts?
Tom: Let's go to Cafe Valencia.
It's right around the corner.
We can sit on the terrace,
above Brighton Square.
Melanie: Ah, that sounds nice.
John: You certainly know your way around
Brighton ...
Waitress: Here's your bill.
Tom: Thanks.
Melanie: How much is it anyway?
Tom: Two pounds seventy.
John: What?
Tom: That's what it says here:
two pounds seventy.
Melanie: Oh, but that can't be right.
John: It must be a mistake.
Tom: We've still got the menu.
Let's see how much it should be.
John: You work it out,
and I'll go and get the waitress.

Look at the bill in your book and listen again.
What's wrong with their bill?

Cassette script

Unit 11 — 2 At the foreign exchange counter

We are at the foreign exchange counter in one of the branch offices of the National Girobank. The first customer is Dennis.

Bank clerk:	Can I help you, sir?
Dennis:	Yes. I'd like to change two hundred German marks.
Bank clerk:	Into Pound Sterling?
Dennis:	Yes, please.
Bank clerk:	That's eighty-one point sixty-three. Two point forty-five German marks to the pound.
Dennis:	Aha.
Bank clerk:	Fifty, seventy, eighty pounds and one, fifty, sixty, three pence.
Dennis:	Thanks very much.

And now Amy Lindner. She is the next customer.

Amy:	Where can I cash travellers cheques, please?
Bank clerk:	Right here, madam.
Amy:	I'd like fifty pounds, then.
Bank clerk:	All right. Sign here, please.
Amy:	(signs)
Bank clerk:	Can I see your passport, please?
Amy:	Yes, here.
Bank clerk:	Right, Ms Lindner. Twenty, forty, fifty. Enjoy your stay in England.
Amy:	Thanks very much.

What does Amy want? Does she get what she wants?
And what about Dennis?
Listen again and check the services in your book.

Unit 11 — 4 Foreign exchange rates

Listen to the exchanges rates on Radio International,
read by Buky Lardner.

And now the exchange rates:
The US dollar is worth one point 73 German marks today,
the British Pound Sterling two point 52,
the Canadian dollar one point 27 today,
and the Australian dollar one mark and nine pfennig.
One hundred Swiss francs will get you one hundred and nine marks,
one hundred Belgian francs four marks 85,
one hundred French francs 29 point 29,
29 Marks 29, what do you know.
But the next one's even better:
one hundred Dutch guilders are worth 88 point 88 marks.
One hundred Austrian schillings 14 point 20,
one hundred Spanish pesetas one mark forty and
one thousand Italian lira will get you one mark and nine pfennig.
One hundred Danish crowns are worth 25 point 76,
one hundred Swedish crowns 22 marks on the button,
one hundred Norwegian crowns 23 point 32,
and one hundred Japanese yen one mark thirty.
The time now is eleven minutes before the hour,
eleven and a half before nine o'clock,
or eight forty-eight and thirty seconds
if you like that better ...

Quite a lot of numbers, eh?
Listen again and take notes in the foreign exchange chart in your book.

Unit 12 — 1 I need stamps

Look at the photo in your book and listen to these dialogues.

Customer:	Good afternoon. Can you help me?
Post office clerk:	Yes?
Customer:	I need stamps for a letter and some postcards, to Germany.
Post office clerk:	Could I have the letter, please?
Customer:	Yes, here.
Post office clerk:	That's twenty-two p, up to twenty grams.
Customer:	Ah yes.
Post office clerk:	And the postcards are the same. How many do you need?
Customer:	Erm ... ten.
Post office clerk:	So ... that's two pounds twenty.
Customer:	Two pounds, twenty pence. Thank you very much.
Tom:	Excuse me. I'd like this birthday card sent special delivery to Germany.
Post office clerk:	Er ... right. That'll be one pound fifty in addition to the normal rate which is twenty-two p ...
Tom:	So, that's one pound seventy-two, right?
Post office clerk:	That's right, sir. Thank you.

Did you understand what Amy, the first customer, needed?
And what does Tom want?
Listen again and answer the questions in your book.

Unit 13 — 3 Pollution

This song was recorded in July 1965 at the legendary Hungry I Club in San Francisco, California.
Words and music by Tom Lehrer.

If you visit American city
You will find it very pretty.
Just two things of which you must beware:
Don't drink the water and don't breathe the air.

Pollution, pollution, they've got smog and sewage and mud:
Turn on your tap, and you get hot and cold running crud.

See the halibuts and the sturgeons
Being wiped out by detergents.
Fish gotta swim and birds gotta fly,
But they don't last long if they try!

Pollution, pollution: You can use the latest toothpaste.
And then rinse your mouth with industrial waste.

Just go out for a breath of air,
And you'll be ready for medicare,
The city streets are really quite a thrill:
If the hoods don't get you, the monoxide will.

Pollution, pollution: wear a gas mask and a veil:
They you can breathe, 's long as you don't inhale.

Lots of things there that you can drink,
But stay away from the kitchen sink:
The breakfast garbage that you throw into the bay
They drink at lunch in San José!
See the crazy people there,
Like lambs to the slaughter
They're drinking the water
and breathing (gulp) the air!

Unit 14 2 A new radiator

Look at the tool box in your book and listen to this scene. Ready?

Peter: What are you doing?
Tina: I'm putting in a new radiator.
I need to take off the hoses first.
Peter: Er ... here's the screwdriver.
Tina: Thanks.
It's no good. I can't pull the hoses off.
Peter: Try hitting it with a hammer. Here.
Tina: Okay.
Peter: Does that work?
Tina: No. I'll have to saw it off.
It needs a new hose anyway.
Can you give me the small spanner
so I can get the radiator off?
Peter: And here's the new radiator.
Tina: Oh no! The holes for the bolts are in
different places.
Peter: Do you need a drill?
Tina: Yes, please.

Which tools do they need?
Listen again and look at the list of tools in your book.

Unit 15 2 What's it made of?

Look at the photos in your book and listen to this scene.
Ready?

Amy: This is about things made of metal, right?
Howard: Well, let's see what we've got?
Kathryn: Hm ... a wrought iron gate.
Tom: Mhm, and a kitchen sink ...
Kathryn: ... made of stainless steel.
Amy: I know that, but what's this?
Howard: A derailleur, for switching gears on a
ten-speed bike.
Amy: How do you know that?
Howard: I've got one ... a ten-speed bike, I mean.
But I don't know what a derailleur is made of.
Tom: It's made of alloy steel, to keep it light weight.
Kathryn: And this seems to be a cylinder block.
Howard: Two, four, six, eight ... sixteen cylinders.
Maybe a ship's engine?
Amy: What's it made of?
Tom: Cast iron alloy?
Kathryn: Could be. This one must be the milling tool,
then.
Howard: Right. And it's made of high speed steel.

Did you understand what the objects are called in English?
And what they are made of?
Listen again and tick the items in your book.

Unit 16 4 Self-tapping or self-drilling?

Look at the drawing in your book and listen to the instructions.
Ready?

Male instructor: Punch or drill a hole through both pieces
of sheet metal. The hole should have the
same diameter as the screw shaft. Place
the tip of a self-tapping screw into the
hole, and drive it in with a screwdriver
or a nutdriver until it is tight.
Female instructor: It's faster if you use a self-drilling
sheet-metal screw with a tip shaped like
a drill bit. This screw drills its own
hole. But you should use a power drill
fitted with a screwdriver or a nutdriver
attachment.

Which type of screws are mentioned?
Listen again and complete the information required.

Unit 17 2 Making a notch

Look at the pictures in your book and listen to the instructions.
Ready?

Female instructor: Measure the lines you later have to cut.
Scribe the cutting lines with a scriber
and repeat this marking procedure for all
the lines you have to cut. Then mark the
lines with a punch and a ball-peen hammer.
Male instructor: Clamp the metal stock in a vice, with the
scribed cutting line running vertically
and close to the jaws of the vice. Place
the saw blade on the waste side of the
line and pull lightly across the metal
until the blade bites into the edge.
Then, using both hands, cut along the
line, applying light, even pressure on the
forward stroke and no pressure on the return.
Female instructor: After cutting the sides of the
notch with the hacksaw, cut the base of
the notch with a flat chisel. Hold the
chisel upright, aligning its blade with
one end of the cutting line and tap
lightly with a ball-peen hammer.
Male instructor: File with smooth forward strokes,
returning the file each time to the
starting position. Smooth one side of the
corner at a time.

Did you get the sequence of the pictures?
Listen again and number them.

Cassette script

Unit 18 3 A machine tool operator

Look at the drawings in your book and listen to this interview. Ready?

Interviewer:	Can I ask you a few questions, sir? About what you are doing?
Operator:	(turns off the machine) Sorry, madam?
Interviewer:	Could you tell us what you're doing at the moment?
Operator:	Well, I'm milling this part into shape.
Interviewer:	So, that's a milling machine you're working on?
Operator:	Right, madam.
Interviewer:	And what is your job?
Operator:	I'm a machine tool operator.
Interviewer:	What does a machine tool operator do?
Operator:	Other than working on milling machines, you mean?
Interviewer:	Yes.
Operator:	I can work on all the machines you see in here, like on the centre-lathe over there, on the surface grinder in the back, and on the drilling presses, of course.
Interviewer:	And where did you learn to operate these machines?
Operator:	During apprentice training on the job. And at the vocational school I went to ...

What is the man doing? What is his job?
Listen again and tick the machines he works on.

Unit 19 3 A pillar drill?

Tina is showing some visitors around the machine shop.
Look at the pictures in your book. Ready?

Tina:	I'll be showing you our drilling machines first. They're all floor models. Right, Peter?
Peter:	Yes, we haven't got any bench models here.
Visitor:	You seem to like your job.
Peter:	Well, after weeks of filing this little pillar drill is a real delight.
Visitor:	A pillar drill? Phh ... sounds Greek to me!
Tina:	Well, most upright drill presses have a pillar, or column.
Peter:	Look here: this is the pillar. It's mounted on the base, and it holds the table for the work pieces to be drilled.
Tina:	And on top of the pillar, that's the drilling head, with the motor.
Visitor:	Good luck with your pillar drill!
Peter:	Thanks.
Visitor:	All I know is the electric drill.
Tina:	Yes, that would be a hand-tool. We've got some of those in our repair shop.
Visitor:	That one's got a pillar as well.
Tina:	Basically, yes. But it's a radial arm drill press.
Visitor:	Because the drilling head is mounted on ... mhm ... this arm here?
Tina:	Right. And now I'd like to show you our machining centres ...

Which of the machines did Tina describe?
Listen again and take notes.

Unit 20 4 Reduce the diameter

Look at the drawing in your book and listen to this dialogue.

Apprentice:	I've got the bar in place. Now what do I do?
Foreman:	First you do the face: Start at the outer edge, and move in towards the centre.
Apprentice:	Okay, the facing.
Foreman:	Hm. Then reduce the diameter to forty-two millimetres, according to your drawing.
Apprentice:	Then the longitudinal turning?
Foreman:	Right. And right at the shoulder you cut a groove leaving a diameter of thirty-six millimetres.
Apprentice:	Aha, the formcut at the end of the thread.
Foreman:	Good. Call me again when your are ready for the threading.

Did you catch the diameters?
Listen again and write them into the drawing.

Unit 21 2 We've got the gutters

Look at the drawing in your book and listen to Susan and Peter. Ready?

Peter:	Let's see what we've got here ...
Susan:	We've got a hundred-and-fifty millimetre gutter.
Peter:	That's capital D for diameter. Nominal size three hundred and thirty-three.
Susan:	Mhm. And in the table we can find all the other dimensions.
Peter:	Small d twenty millimetres.
Susan:	That's the gutter roll diameter.
Peter:	But we've got the gutters. What we need are the brackets.
Susan:	Yes. Look here. We need brackets with two springs.
Peter:	Small d one is a hundred and fifty-three millimetres, of course.
Susan:	How long's the arm of the bracket?
Peter:	That's ... c two.
Susan:	Two hundred and thirty. And the long ones three hundred.
Peter:	And what about the profile?
Susan:	There are three choices. Thirty by five millimetres should do.
Peter:	Profile number one then. And what's ... what's d two?
Susan:	That must be the hole for fastening the brackets.
Peter:	Ah here. For s up to five millimetres d two is six millimetres.
Susan:	Sounds good.

Well, what parts are they talking about? And what size?
Listen again and mark the part and size they need.

Unit 22 **2 The functions**

Look at the computer components in your book and listen. Ready?

Trainee: I know about the basic hardware:
the central processor, the monitor, keyboard, graphics tablet, the mouse, printer or plotter and scanner.
But what do the single components do?
Engineer: The computer is the heart of the system.
It does all the calculations.
On the monitor you can see what you draw and you can see what the computer wants to tell you.
The keyboard and graphics tablet are used to feed the computer the necessary information.
If you want to see the result of your work on paper, you need a printer or a plotter.
Trainee: What do you need the scanner for?
Engineer: With the help of the scanner
you can translate a drawing on to paper
so that the computer can understand it.
Trainee: That sounds easy.
So where are the difficulties?
Engineer: There are difficulties only in the beginning.
You have to learn a large number of commands which the CAD programmes require.
Once you've learned how to communicate
with the computer you'll enjoy working it.
Trainee: Well, I'll try my best.

What did you learn about CAD?
Listen again and connect the components and their functions.

Unit 23 **1 Like kangaroo and elephant?**

Look at the picture in your book and listen to the dialogue. Ready?

Visitor: What is it supposed to be?
Owner: A Kangofant.
Visitor: A what?!
Owner: A KANGOFANT.
Visitor: Like kangaroo and elephant?
Owner: Right!
Visitor: But what is it?
Owner: A tennis ball collecting machine.
Visitor: Collecting tennis balls?
Owner: Yes. The kangofant collects the balls for you on the tennis court.
Visitor: How does it do that?
Owner: Wall boards are installed at one end of the court. You play the balls.
From the wall board they're deflected
on to a screw conveyer.
They are then fed to a blower,
and blown through a hose to the container.
There you pick up your new balls.
Visitor: Now I understand.

Do you know what a KANGOFANT is?

Unit 24 **1 A lot of waste**

Interviewer: As consumers we ... we create a lot of waste.
A household of four people throws away about one point three tons of rubbish a year.
We buy products, use them
and throw them away.
Expert: I know.
And we don't want to deal with the rubbish.
We want it as far away as possible.
Interviewer: So what do we do?
Expert: Well, we can bury waste in landfills,
burn it in incinerators or dump it at sea.
Interviewer: In Germany?
Expert: Well, yes. We export it, and some other country will dump it for us.
Interviewer: So, where is the problem?
Expert: Well, take landfills for example.
Who wants to live near one?
First, there is the smell.
And toxic chemicals can leak out
and get into the ground water.
Interviewer: What about burning the waste?
Expert: The problem is that waste does not disappear.
When you burn three tons of rubbish,
you produce one ton of ash containing heavy metals.
Interviewer: And dumping into the sea is no solution either?
Expert: Of course not!
Every year more than two million seabirds
and over one hundred thousand whales
and other marine mammals die.
They die because they become entangled in plastic debris or because they eat it.
Interviewer: It seems that all we can do
is to reduce, reuse and recycle.
Expert: How very true.

Listen again and answer the questions in your book.

Unit 25 **1 Welding is dangerous**

Visitor: Can you tell me something about the welding techniques used in your company?
John: Yes, er ... hm.
In our factory we use four different types of welding equipment: arc welding, oxyacetylene gas welding, MIG welding and spot welding.
Vistitor: Are there any risks involved?
John: Oh yes, welding can be dangerous
if you don't know what you're doing.
You have to wear heatproof gloves and a mask to protect your face and eyes.
Visitor: Er ... what's this in the corner?
John: It's an oxyacetylene set.
Visitor: Those two cylinders look like torpedos!
John: Aha. If you don't handle them with care
they can go off like torpedos!
Visitor: And ... what are they for?
John: The tall one contains the oxygen and
the shorter one contains the acetylene
and, as you can see, they are chained together on a two-wheeled handcart.
Visitor: And how do you get to the gas?
John: You can open the valves of both cylinders with these spindle keys.

Visitor: Aha. And there you can regulate the pressure?
John: Oh yes. On each valve a dual regulator with two gauges measures the pressure in the cylinder and the working pressure in the hose. The working pressure is adjusted by means of this knob on the front of each regulator ...
Visitor: What are the red and blue hoses for?
John: Er ... the red one is for acetylene and the blue or black for oxygen. They carry the gases to the torch, where an oxygen valve and acetylene valve control the proportion of each gas at the tip of the torch.

Which types of welding are used in the company where John works?
Listen again and tick the types of welding.

Unit 26 — 2 What is MIG welding?

Look at the photos in your book and listen to the interview.
Which of the photos are they talking about?

Interviewer: I'm talking to John.
He is a MIG welder in a large factory.
John, what is MIG welding?
John: MIG welding is a wire-feed arc welding process and MIG stands for metal inert gas.
Interviewer: Inert gas? What is that for?
John: The gas is needed to protect the arc and the molten weld metal from oxygen.
Interviewer: Which gas do you use?
John: Oh, that depends on the metal we're welding.
Interviewer: So, you use different gases?
John: Oh yes. For stainless steel we use argon with a small amount of oxygen added.
For copper we use helium or an argon-helium mixture.
And for aluminium we use pure argon or carbon dioxide.
Or a mixture of argon and carbon dioxide.
Interviewer: And why wire-feed?
John: The filler wire is fed through the welding gun by a wire feeder.
Interviewer: Like an electrode from a wire-reel?
John: Right.
Interviewer: And where do you get the power from?
John: From the power source which provides currents up to 330 amps.
Interviewer: Aha. And where do you use MIG welding?
John: Oh ... we use it for maintenance and repair work. For joining sheet metal, tubes and angles or box sections.
But it can also be used in production processes. For a wide range of metals, like low carbon steel, or stainless steel, for aluminium and magnesium alloys, and for copper.
Interviewer: And is it hard to learn?
John: Well ... it took me a few days to ... to do simple welds.
But quality welding takes quite some practice.
I'm working on it.
Interviewer: Well, good luck then.

What are they talking about?
Listen again and answer the questions in your book.

Unit 27 — 1 Number one

Robot: Look at me and listen.
I am number one welding robot.
Ask me any question you want.
Welder: What are you working on?
Robot: It is a bucket for a ten ton earth moving device.
Welder: A caterpillar, you mean?
Robot: Yes.
Welder: How do you move it?
Robot: By the workpiece rotating device.
Welder: How do you know which way to turn it?
Robot: Programme number one zero zero one.
Welder: How do you know what to weld and where?
Robot: What or where?
Welder: What do you mean?
Robot: Only one question at a time, please.
Welder: How do you know what to weld?
Robot: Programme number two zero zero two.
Welder: And where?
Robot: Programme number three zero zero three.
Welder: So you are always on the right spot?
Robot: Right!
Welder: And what do I do?
Robot: You feed me with the programmes.

Do you understand what they are talking about?
Listen again and tick the answers in your book.

Unit 28 — 2 The silicon rubber mold

Customer: I've read about your spin-casting in a professional magazine.
And I'd like some more information.
Technician: Aha. Go right ahead.
Customer: We produce small plastic parts such as cog-wheels and other components for lawn-mowers.
Technician: So, you've used plastic injection molding?
Customer: Right. But our molds are too expensive for smaller prototype series.
Technician: Yes. That's were spin-casting is a lot more cost-effective.
Customer: How much would a ... a typical mold for, say, six cog-wheels cost?
Technician: Well, we have set up molds for about a hundred dollars.
Customer: Is that the silicon rubber mold?
Technician: Yes.
Customer: How long would such a mold last?
Technician: About two hundred casts.
Customer: Very interesting.
Can I send you a few of our parts for a sample cast?
Technician: Yes, sure.
We can run off the cast in a few hours.
And you'll have our samples within a few days.
Customer: Sounds good.
Technician: Can I have your name and address, please?

Do you understand what they are talking about?
Listen again and tick the answers in your book.

Unit 29 3 Changing the mould

Look at the photos in your book and listen to this conversation.
Ready?

Visitor:	Excuse me! Can I ask you what you are doing here?
Machine operator:	Changing the moulds.
Visitor:	Changing the moulds?
Machine operator:	Yes. I'm supposed to run off a batch of shaver housings now.
Visitor:	And before that you did those telephone keys?
Machine operator:	Mhm.
Visitor:	The mould has two parts?
Machine operator:	Yes. They're mounted semi-automatically, like this.
Visitor:	Wow!
Machine operator:	And now we shut her tight with the clamping unit. Here we are. Ready to go.

What is the man doing?
Which parts are being made?

Unit 30 4 Planetary grinding

Look at the drawing in your book and listen to this conservation.
Ready?

Visitor:	It says here that the facing head can be used for planetary grinding. What do you mean by that?
Technician:	Well, normally simple cylindrical components can be rotated against the grinding tool ...
Visitor:	Which is rotating as well?
Technician:	Right. The tool and the component are rotating.
Visitor:	And planetary grinding?
Technician:	We use planetary grinding for large components that can not be rotated.
Visitor:	So, the component is stationary?
Technician:	Yes. The grinding tool, which is rotating at a high speed, moves around the surface of the component.
Visitor:	I see. So planetary means that the tool moves around the component like a planet.
Technician:	Right.

Do you now understand planetary grinding?
What is stationary and what is being moved?
Listen again.

Unit 31 3 Safety gear

Foreman:	Hey buddy! Stop!
Jack:	What?!
Foreman:	Not allowed in without a hard hat!
Jack:	I can't stand it. Gives me a headache.
Foreman:	Would you rather have your head bashed in? Or your toes cut off?
Jack:	Eh?
Foreman:	You need to wear safety shoes here.
Jack:	You mean I can't work like this?!
Foreman:	No way! A steel mill is not a tea party.
Jack:	I thought ...
Foreman:	We've got more 'n a thousand degrees at the furnace. And fireworks flying all over the place when we tap it.
Jack:	The furnace?
Foreman:	Right. So better get yourself a leather apron, a face shield ...
Jack:	Safety shoes, hard hat and gloves, right?
Foreman:	Right, and ear muffs. I see you're catching on.

They're talking about workwear, aren't they?
Listen again for the items they mention.

Unit 32 3 Chain's a bit dry

Tina is looking for a motor bike.
Look at the checklist in your book and listen to this dialogue.
Ready?

Tina:	Looks okay to me.
Salesman:	Yes. It's been thoroughly checked and serviced.
Tina:	Well, the chain's a bit dry.
Salesman:	That's because it's been here for a while.
Tina:	And the rear wheel bearing has a bit of play.
Salesman:	Hm. Probably needs a bit of grease.
Tina:	There's only about five hundred miles left on the rear tyre.
Salesman:	I'll see what we can do about that.
Tina:	And the front brake needs adjusting.
Salesman:	Yes. Yes, we can do that right away.
Tina:	Can I sit on it?
Salesman:	Certainly.
Tina:	Clutch and front brake feel all right.
Salesman:	And the electrics should be in order.
Tina:	The brake light is okay, headlight also, and the indicators ... (hoots)
Salesman:	And the horn.
Tina:	Can I test-drive it?
Salesman:	Certainly.

Tina (starts engine, drives off).

Which parts of the checklist did they discuss?
Listen again and tick the items in your book.

Cassette script

Unit 33 **2 It's got 100 hp**

Look at the pictures of the old and the new tractor in your book and listen to this conversation. Ready?

Buky: There's a tractor for you!
Dennis: Oh ... when do you think that was built?
Buky: Around nineteen-twenty.
Howard: It doesn't look that much different from today's tractors.
Dennis: And it pulls three ploughs.
Buky: That's where you've got one difference: The eighty-one is a lot more powerful.
Dennis: Yes. It's got seventy-three point five kilowatts.
Buky: That's one hundred hp. And the "Allwork"?
Dennis: Twenty-five hp ...
Buky: That's at the belt. At the drawbar, where you need it, it's only twelve hp.
Dennis: I see. The engine delivers twenty-five hp, and most of that is used up in the transmission ...
Buky: By the transmission belt.
Howard: But the "Allwork" is lighter than the 8120.

What do you think of the two tractors?
Listen again and mark the data they are talking about.

Unit 34 **3 Catalytic converters**

Look at the drawing of the three-way catalyst system in your book and listen to this interview.
Ready?

Interviewer: Er ... everybody is talking about car pollution and how to prevent it. Can you tell me how it works?
Expert: Well ... the major component is the catalytic converter.
Interviewer: Yes, I've heard that. How does it work?
Expert: The principle is simple: an active substance, the catalyst, causes a chemical reaction that turns exhaust emissions into harmless substances.
Interviewer: Entirely harmless?
Expert: Well, relatively harmless.
Interviewer: So the whole thing is more complicated?
Expert: Mhm yes. Converters consist of two sections. The front section is called a three-way catalyst because it turns hydrocarbons, carbon monoxide and nitrogen oxides into water, carbon dioxide and nitrogen.
Interviewer: But carbon dioxide is responsible for the greenhouse effect!
Expert: Yes. The catalytic converter is not a final solution. It only reduces the exhaust emission under certain conditions.
Interviewer: Which are?
Expert: Well, you need unleaded fuel, and the air-fuel mixture must be kept at 14 point six to one.
Interviewer: How can this be done?
Expert: The carburettor must be computer-controlled.
Interviewer: Sounds complicated.
Expert: Shall I go on?
Interviewer: Yes, please do.
Expert: The rear section of the catalytic converter is an oxidation catalyst which further reduces the hydrocarbons and carbon monoxide in case the air-fuel ratio is off.
Interviewer: And that's it?
Expert: No. There is also an air pump which supplies air to oxidize unburned gasoline in the exhaust. And it provides the oxygen needed for the oxidation catalyst.
Interviewer: Thank you very much. Let's hope catalytic converters will slow down the greenhouse effect.

Sounds confusing, huh?
Listen again and tick the words in the drawing.

Unit 35 **4 Signals from the sensors**

Look at the engine drawing in your book and listen to this interview.
Ready?

Expert: If you look at this schematic drawing you can see most of the components of a computer-controlled engine.
Interviewer: Well ... I can see the computer.
Expert: Right. The computer receives signals from sensors. Take the crankshaft position for instance: The computer must know the position of each piston, so that it can trigger the electronic ignition at the right moment.
Interviewer: I see. And the sensor tells the computer where the pistons are?
Expert: Indirectly, yes. The sensor monitors the position of the crankshaft.
Interviewer: Aha. But there are many more sensors.
Expert: Yes. There are sensors that monitor throttle position, the amount of oxygen in the exhaust gas, coolant temperature etcetera.
Interviewer: And what does the computer do with all these data?
Expert: Based on these data the computer sends commands to the ignition system and to the other components.
Interviewer: To the operating hardware?
Expert: Yes, mostly small motors or solenoids and valves.
Interviewer: Mhm. And what does EGR stand for?
Expert: Exhaust gas recirculation. That's important for pollution control.

Do you see the components they are discussing?
Listen again and tick the components they mention.

Unit 36 4 Rim Rider

Look at the photos in your book and listen to this presentation. Ready?

Speaker 1: Rim Rider is super-light, low-noise and environmentally friendly.
Speaker 2: Gone are all major metal parts.
Even the engine block and frame are made of carbon fibre and space-age plastics.
Speaker 1: Gone are chain and belt drive systems.
A hydraulic system drives a cogwheel inside the rear wheel.
Speaker 2: Gone are axles, spokes and centre hubs.
The front rim is solidly fixed to the front swing arm.
And the tire revolves around the stationary rim.
Speaker 1: Gone are air-filled tires.
The tires are solid and can be filled with gel to help suspension.
Speaker 2: Gone is the muffler pipe.
Noise is reduced to almost zero by an electronic synthesizer.
Speaker 1: Additional features include ABS front brakes, hydraulic steering, an automatic transmission and an air-cooled fuel-injected two-stroke engine from 125 to 350 cc.
Speaker 2: The Rim Rider is the dirt bike of the future.

May well be. Did you find out which product they presented?
Listen again for the features of this futuristic enviro-bike.

Unit 37 2 A useful application

Look at the photos in your book and listen to the interview with Mr Devol, a robot expert. Ready?

Interviewer: Everyone dreams of having a robot to do the housework.
What do you think of that, Mr Devol?
Devol: Well, cleaning up or cooking seem to be simple tasks.
But they consist of so many different movements, that robots would require millions of instructions.
Interviewer: What's this then?
Devol: That's more like an electronic toy, that can bring the paper or the mail.
Interviewer: But how can robots help people?
Devol: Other than in factories?
Interviewer: Yes.
Devol: Well, here we have an interesting development for blind people.
This robot is programmed with a map and it's fitted with ultrasonic sensors.
Interviewer: Like a robot "guide dog"?
Devol: Well, yes, but not as good as a dog.
Interviewer: And what's happening here?
Devol: This is a workplace for a disabled person who works with the help of a robot.
Interviewer: That looks like a very useful application.
Devol: Yes. And then there are wheel-chairs fitted with small robot arms to help disabled people.
We also have library chairs which are ...

Which of the pictures are they talking about?

Unit 38 2 Redundant in six weeks' time

Interviewer: I've heard that the factory will be closed in the near future.
Is it true?
Ray Ede: Yes, that's correct. This is part of ... an STC rationalisation plan.
The factory is in fact being closed and consequently all of the people would be redundant in about six weeks' time.
Interviewer: Oh! How long have you been working here?
Ray Ede: I've been here for nearly twenty years now.
Interviewer: So how do you feel about this?
Ray Ede: An, mixed feelings I guess!
Very disappointed to see the amount of expertise lost, and the amount of good people that we've built up over a long period of time ...
Interviewer: Yeah, I see ...
Ray Ede: We are now, I feel, at the peak of our production capability and to see that capability going is ... is a bad thing.
Interviewer: I can imagine.
Ray Ede: It's ... but I think everybody is ... is quite philosophical about it, because it's a fact of life:
Companies are closing ... not just in this country but ... but all over the world.
If I tell you that ... five years ago it used to take us fifty man hours to produce one teleprinter.
It now takes us six hours because of automated processes.
So if you follow that through it means, that if we are using the same number of people, we have to sell nearly ten times as many machines.
Interviewer: I see.
Ray Ede: Or: if we sell the same amount of machines, we can do it with a tenth of the labour force.
And this is one of the ... drawbacks, I guess, of modern technology.

So, hi-tech is not all good.
Look at the statements in your book and listen again.

Unit 39 2 Testing prototypes

Barry Brindle: Basically, my job is to plan and run series of tests on prototype machines to make sure that the engines are capable of the work the equipment demands. Our engines go into many types of vehicles: tractors, combine harvesters, forklift trucks and so on.
When customers come to us they specify the operating conditions for the engine they need. So when the engine team has come up with a prototype, my job is to test it to make sure it satisfies both the customer's and our own specifications.

What is the man talking about?
Listen again and tick the answers in your book.

Cassette script

Unit 40 **2 Job scene**

Caller: Eight, oh, seven, seven.
Alice Moss: Hello, caller!
This is Brighton Job scene, bringing you details of some of the jobs on offer in the Brighton area.
If you are interested in any of them write down the reference number and ring me: Alice Moss, at the Job Centre, that's Brighton 69 35 double 9.
Man: First we are looking for a skilled operator to work with new machinery in a large pipe factory.
You should have finished your schooling and passed your City and Guilds exams but you don't need any experience.
The company will give you full training.
The pay at the beginning will be modest but you will have many benefits and excellent prospects for the future.
The reference number is Brighton 3-8-1-6.
Alice Moss: Next we have job opportunities for bodywork and repair mechanics at a garage in Worthing.
You should have some experience, be self-reliant and be able to make decisions on your own.
The expanding firm offers you the latest equipment and some training in new methods of bodywork.
You will be paid according to your experience.
The reference number here is Brighton 2-2-4-4.
Man: And finally we are looking for engineering technicians in the field of ...

Sorry, but we have to leave Job Scene.
Did you understand what the jobs were about?
Listen again and take notes in your book.

Unit 41 **3 Schools in England**

Amy: What's education like in England, Howard? Would you tell us?
Howard: Well, children start off at nursery school, up to the age of five.
Amy: Oh. They start quite young.
Howard: Nursery school isn't compulsory, but it helps young children to fit into school more easily.
Amy: Mhm.
Howard: Then they go to infant school until they are seven, and then on to primary school until they are eleven.
Amy: And what do they do after primary school?
Howard: Well, most of the children go off to secondary school, for GCSE.
Amy: What's that?
Howard: General Certificate of Secondary Education. They take that when they're sixteen.
Amy: And then?
Howard: And then they either go on in secondary school until they are eighteen and take their A'levels.
Or they can go off to vocational school or technical college.
Amy: I see.
And what about your own schooling, Howard?
Howard: Ha. Well, that was different ...

Listen again to the cassette and look at the diagram in your book.
Fill in the ages and tick the schools Howard mentions.

Unit 42 **5 Get organised**

Anne Moss. This is Anne Moss from Brighton Job Scene, today with some tactics to plan your job hunting campaign.
Man: First of all get organised.
Get a small notebook to put down the names of the firms you want to write to.
Put the date you write, the date you get a reply and wheter it is a 'No' or 'Contact us later'.
Always follow up a letter or application form if you have had no reply after two or three weeks.
Anne Moss: Learn how to write letters, fill in application forms and prepare for interviews.
Take help and advice and try, try again.
It's not just a case of getting a job, but getting the job you want.

Unit 43 **1 Can you make it at nine?**

Ms Bainbridge: Hello?
Mr King: Good morning.
Is this Catherine Bainbridge speaking?
Ms Bainbridge: Yes. Catherine Bainbridge. Good morning.
Mr King: This is Martin King from Tectex Brazil.
I'm in London for two weeks.
Ms Bainbridge: Oh yes. Mr King.
Mr King: I'm calling about your application as a machine fitter.
We're interviewing next Monday.
Can you make it at nine a.m.?
Ms Bainbridge: I'd be delighted.
Mr King: You understand we've got quite a number of applicants.
Ms Bainbridge: Yes, I understand.
So, that's Monday, the fourth of June, at nine a.m., at the company.
Mr King: Right, at Ealing, St Mary's Road.
Ms Bainbridge: I've got the address.
Mr King: We'll see you at the Manager's office on the first floor.
Ms Bainbridge: Thank you very much, Mr King.
Mr King: Thank you, Ms Bainbridge.
Ms Bainbridge: Good bye.
Mr King: Good bye.

Unit 44 1 Some international experience

Martin Marek has made an appointment with Ann White of C. Dugard Limited, who is visiting Nuremberg.
At this time, they are meeting in the lounge of the Hotel Europa, and we fade into the interview.

Ms White:	Have you had any serious illness or injuries?
Mr Marek:	No, nothing but the usual measles and a flu now and then.
Ms White:	You look fit to me.
Mr Marek:	Oh thank you.
Ms White:	Where did you go to school? Here in Nuremberg?
Mr Marek:	Yes, I went to primary and secondary school until I was 15. And then I started an apprenticeship.
Ms White:	How does that work in Germany?
Mr Marek:	Well, I work at Haas four days a week, and on Wednesday I attend the Vocational School for machine fitters in Nuremberg.
Ms White:	How long have you been doing that?
Mr Marek:	I started four years ago and I'll have my Diploma at the end of June.
Ms White:	Will you have to take an exam?
Mr Marek:	Yes, something like the City and Guilds examination in England.
Ms White:	How do you know about the City and Guilds certificate?
Mr Marek:	We learned that in English. Why? Is anything wrong?
Ms White:	No, no, not at all. Quite the contrary. Now, why do you want to come to England and work for us?
Mr Marek:	Well, I'd like to get some international experience. And I'd like to work on my English.
Ms White:	But your English is very good.
Mr Marek:	Thank you very much, Ms White.

Unit 45 The Jack of Hearts

We were looking for a new receptionist,
we had three applicants for the job,
they all looked very nice.
Among other things we wanted to be sure
that they could count properly.
This is what happened.

Interviewer:	Can you count, madam?
First applicant:	Certainly: 2-4-6-8-ten-9-7-5-3-1.
Interviewer:	Quite interesting. Where did you learn this?
First applicant:	Well, I used to deliver the mail. So I counted the numbers up on the right hand side of the street, and back down on the left.
Interviewer:	I see.

Our next applicant was a young man.

Interviewer:	Sir, can you count to ten?
Second applicant:	Can I do it the other way round?
Interviewer:	Why not?
Second applicant:	Ten-9-8-7-6-5-4-3-2-1-Zero.
Interviewer:	Very good, but tell me: Why did you count down?
Second applicant:	I used to work for NASA, and I counted down all the missile starts.

The third applicant looked very elegant and he must have heard what we'd asked the others.

Interviewer:	I'd like to ask you to ...
Third applicant:	One-2-3-4-5-6-7-8-9-ten.
Interviewer:	Very good, sir. Can you gon on?
Third applicant:	Jack, Queen, King, Ace!

Word List

The **Word List** only contains such words which we think you might want to look up. It only includes the one (or two) meaning(s) in German that the English word listed has in context. The **Word List** should help you as a quick and handy reference whenever you want to check a word or expression and when you haven't got any dictionary within reach. It cannot and shall not be substituted for serious work with a **dictionary.**

We have used the following **abbreviations:**

abb	abbreviation
fr	French (aus dem Französischen)
GB	British English
span	Spanish
US	American English

A

abbreviation	Abkürzung
abide by	sich halten an
ability	Fähigkeit
abolish	abschaffen
above	über, oberhalb
abrasive	abschleifend
abrasive wheel	Schleifscheibe
abroad	im (ins) Ausland
accelerate	beschleunigen
accept	akzeptieren
access	Zugang
accessibility	Zugänglichkeit
accessible	leicht zugänglich
accessory	Zubehör
accommodation	Unterkunft
accompany	begleiten
account	Konto
acetylene	Acetylen
achieve	erreichen
across	über
active	rührig, wirksam
activity	Tätigkeit
actress	Schauspielerin
actuator	Verstellorgan
acute	spitz
adapt	anpassen, einfügen
adapter	Steckhülse
add	hinzufügen, addieren
addition: in addition to	Addition: zusätzlich zu
additional	zusätzlich
adept	erfahren
adjust	regulieren
adjustable	regulierbar
adjusting knob	Einstell-, Regulierknopf
adjustment	Einstellung
advanced	fortschrittlich, modern
advantage	Vorteil
advertisement	(Zeitungs-) Anzeige, Inserat
advice	Rat
affect	beeinflussen
afford	sich leisten können
affordable	erschwinglich
agree	zustimmen
ahead	voraus
aid	helfen; Hilfe
air-cooled	luftgekühlt
air-fuel mixture	Luft-Kraftstoffgemisch
alert	lebhaft, aufmerksam
align	ausrichten, eichen, trimmen
all-purpose wrench	Kombizange
allowance	Spesen, Vergütung
alloy	Legierung
alloy steel	legierter Stahl
alteration	Änderung
ambitious	ehrgeizig
ammeter	Amperemeter
amorphous	formlos
amount	Menge, Betrag
amps (abb) ampere	Ampere
analyze	analysieren
anchor	verankern
anchovy	Sardelle
angle	Winkel
angular	winklig, eckig
anticlockwise	gegen den Uhrzeigersinn
antifreeze	Frostschutzmittel
apostrophe	Apostroph
appd (abb) approved	genehmigt
applicant	Bewerber/in
application	Anwendung, Bewerbung
application form	Bewerbungsbogen
apply	anlegen, anwenden; sich bewerben
appointment	Verabredung, Treffen
appreciation	Verständnis
apprentice	Lehrling, Auszubildende/r
apprenticeship	Lehre
appropriate	passend
approved	anerkannt, bewährt; genehmigt
approximate	ungefähr
apron	Schloßplatte (Drehmaschine)
arbour	Welle
arc	Lichtbogen
arc welding	Lichtbogenschweißen
area	Fläche, Areal
arrangement	Anordnung, Vereinbarung
arrival	Ankunft
arrive	ankommen
ash	Asche
assemble	zusammenbauen, montieren
assembly	Montage, Fertigung
assembly-line worker	Fließbandarbeiter/in
assess	einschätzen
assist	helfen, unterstützen
associate	in Verbindung bringen
assort	(an-)passen
assurance	Zusicherung
atmospheric pressure	Luftdruck
attach	befestigen, anhängen, beifügen
attachment	Verbindung, Befestigung, Zusatzgerät
attempt	versuchen, anpacken
attend	(Schule) besuchen
attention	Aufmerksamkeit
attract	reizen, anziehen
augment	vermehren, zunehmen
authorities	Verwaltung, Obrigkeit
authority	Autorität, Einfluß
available	vorhanden, erhältlich
avoid	vermeiden
award	Auszeichnung, Preis
awareness	Bewußtsein
axis	Achse (geometrisch)
axle	Achse
axle load	Achsgewicht

B

back	Heck, Rückseite
backbone	Rückgrat
backup	auftragen; Unterstützung
bacon	Speck
baked	gebacken
baking	brennen, sintern
ball-bearing	Kugellager
ball-peen (pein) hammer	Kugelhammer
ballast-weight	Zuladung
bar	Stab, Stange

barbecue	Grill
barmaid	Barfrau
base	Grundplatte, Sockel
basic	grundlegend, fundamental
basically	grundsätzlich, im wesentlichen
batch	Menge, Los
batter	Teig
bay	Bucht
bead	Kügelchen
bear	tragen; lagern
bearing	Lager
because	weil
bed	(Maschinen-)Bett; bedecken, abdecken
beef	Rindfleisch
belong	gehören, zugehören
below	darunter
belt	Riemen
bench model	Tischmodell
bend	Kurve; biegen
benefit	Vorteil, Nutzen
bent	gebogen
between	zwischen
beware	sich in acht nehmen
bike	Motorrad
bill	Rechnung
bit	Bohrerspitze, Schneide
blade	Blatt, Klinge
blend (with)	vermischen, ineinander übergehen
blow	Schlag; blasen
blow dry	fönen
blower	Gebläse
blown	durchgebrannt
blue cheese	Gorgonzola
board	einsteigen in
body shell	Karosserie
bodyshop	Spenglerei
bodywork	Karosseriereparatur
boiler	Dampfkessel
bolt	Bolzen; schrauben
bond	zusammenhalten
book	buchen, eintragen
boot	(Gummi-)Manschette
border	Grenze
bore	Bohrung, innerer Zylinder-durchmesser
boring	Bohren, Bohrung
born	geboren
both	beide
bottle bank	Flaschensammelbehälter
bottom	Boden, Grund; tiefster Teil; unten
box	Schachtel, Behälter
bracket	Rinnenhaken
brain	Gehirn
brake	Bremse; bremsen
branch office	Zweigstelle
brand	Marke
brass	Messing
break	Zwischenraum; Pause
breast	Brust
breathe	atmen
bricklayer	Maurer/in
brief	kurz
bright	intelligent; hell
broadbased	weitgefächert
brochure	Broschüre
brought up	erzogen, aufgezogen
bubble	Blase
bucket	Eimer
builder	Erbauer/in, Bauarbeiter/in
building	Gebäude
built-in	eingebaut
bun	Rosinenbrötchen
bury	vergraben
business card	Geschäfts-, Visitenkarte
butcher	Fleischer/in
button	Knopf
by means of	durch, mittels

C

cable	Kabel, Seil, Fahrdraht
CAD (abb) computer-aided design	computerunterstütztes Zeichnen
cake	Kuchen
calculation	Rechnung
call	rufen
calliper	Greifzirkel, Taster
campaign	Kampagne
camshaft	Nockenwelle
cancel	entwerten, absagen
cap	Kopf, (Verschluß-)Kappe
cap nut	Überwurfmutter
cap screw	Kopfschraube
capability	Fähigkeit
capable	geeignet
capable of	in der Lage sein
capacity	Kapazität, Volumen
capital	Großbuchstabe; Hauptstadt
car licence	Zulassungsschein
caraway	Kümmel
carbon	Kohlenstoff, Kohle
carbon fibre	Kohlefaser
carbon monoxide	Kohlenmonoxid
carburetor (US)	Vergaser
carburettor (GB)	Vergaser
care	Sorge, Aufmerksamkeit; sich sorgen
career	Karriere, (Lebens-)Beruf
careful	genau, vorsichtig
carpenter	Zimmermann
carriage	Support
cart	Karren
case	Behälter
cash	Barzahlung, Bargeld; einlösen
cash card	Bankomatkarte
cast	Guß
cast iron alloy	Gußeisenlegierung
catalyst	Katalysator
catalytic	katalytisch
category	Ordnung, Klasse
cater	verpflegen, sorgen
catsup (US)	Ketchup
cause	Ursache
cavity	Hohlraum
cc (abb) cubic centimetre	Kubikzentimeter (Hubraum)
centre	Mitte; zentrieren
centre lathe	Drehmaschine
centre punch	Körner
certainly	sicher(lich), gewiß
certificate	Zeugnis
chain	Kette
chairman, chairperson	Vorsitzende/r
challenge	Herausforderung
chamber	Hohlraum, Kammer
chamfer	Fase, Abschrägung
chancer	Gelegenheitsarbeiter/in
change	tauschen, verändern; Veränderung
channel	Kanal
charge	in Rechnung stellen, Ladung
chargrilled	vom Holzkohlengrill
chart	Plan, Diagramm, Tabelle
chd (abb) checked	geprüft
cheap	billig
check	prüfen; Prüfung, Kontrolle
chef	Küchenchef/in
chemical	Chemikalie
cheque	Scheck
cheque book	Scheckheft
chipper	Behauer, Meißler/in
chipping action	Zerspanungsvorgang
chips	Splitter, Scherben, Späne

Word List

chisel	Meißel
choice	Wahl
choke	ersticken
choose	wählen
chop	hacken
chuck	Spannfutter; Bohr(maschinen)futter auf- und einspannen
circle	Kreis
circlip	Sicherungsring
circuit	Schaltkreis, Stromkreis
clamp	Zwinge; festspannen
clamping levers	Spannhebel
clamping unit	Schließeinheit
classmates	Klassenkollege/in
clean	reinigen; sauber
clearance	Bodenfreiheit
clerical work	Büroarbeit
clerk	Büroangestellte/r
clockwise	im Uhrzeigersinn
close	schließen; genau, dicht, fest
close fit	enge Paßform
cloth	Tuch, Lappen
clutch	Kupplung
CNC (abb) computer numerical control	Steuerung mit Mikrocomputer
coach bolt	Flachrundschraube (Torbandschraube)
coach screw	Sechskantholzschraube
coast	Küste
coat	Anstrich; an-, überstreichen
coated	überzogen
cog-wheel	Zahnrad
coil	Rolle
colleague	Kollege/in
collect	sammeln, einsammeln
collection	Sammlung
college	höhere Schule
colour	Farbe
column	Säule, Ständer
combination	Verbindung
combination pliers	Kombizange
combination set	Winkelschmiege
combine	verbinden, zusammensetzen
combine harvester	Mähdrescher
combustible	brennbar
combustion	Verbrennung
command	befehlen, beanspruchen, Befehl
comment	Kommentar
commercial correspondence	kaufmännischer Schriftverkehr
common	gebräuchlich
common bolt nut	Sechskantmutter
communicate	kommunizieren
company	Firma
compare	vergleichen
compile	zusammenstellen
complete	vervollständigen
component	(Bestand)Teil
compound	Mischung, Masse
compound rest	Oberschlitten
comprehensive	umfassend
compress	komprimieren, verdichten
compression	Verdichtung
computer-controlled	computergesteuert
concentrate	konzentrieren
concentricity	Konzentrizität
concern	betreffen
condition	Bedingung
cone	Tüte
confidence	Vertrauen
confident	selbstbewußt, zuversichtlich
confidential	vertraulich
configuration plan	Konstruktionszeichnung
configure	zusammensetzen, anordnen
confirm	bestätigen
confuse	vermengen, verwirren
conjunction	Verbindung, Zusammentreffen

connect	verbinden
connection	Verbindung
conscious	bewußt
consider	beachten, erwägen
considerable	beträchtlich
consist	bestehen
construct	entwerfen, konstruieren
consultant	Berater
consumption	Verbrauch
contactor (switch)	Kontaktschalter
contain	enthalten
container	Behälter
contamination	Verunreinigung, Verseuchung
contrary	entgegengesetzt
content	zufrieden, Inhalt
continue	fortsetzen
continuous	kontinuierlich, fortlaufend, ununterbrochen
contour	Kontur, Umriß
contract	Vertrag
contracting company	Vertragsgesellschaft
control	kontrollieren
control panel	Schalttafel
controller	Regler
convenient	passend, bequem
conventional	herkömmlich
conversation	Unterhaltung
converse	reden, sich unterhalten
conversion	Umwandlung, Neugestaltung
convert	umwandeln
converter	Umformer
convey	befördern, transportieren
cook	Koch/Köchin
coolant	Kühlflüssigkeit
copper	Kupfer
core	Kern
cored	mit Kern
corned beef	gekochtes und eingesalzenes Rindfleisch
correct	richtig
correspond	übereinstimmen, entsprechen
cost estimate	Kostenschätzung
cost-effective	kostengünstig
couchette	Liegewagen
council	(Gemeinde-)rat
counter	Schalter
counter assistant	Verkaufsassistent/in
countersunk	versenkt
coupling	Kopplung, Kupplung
course	Strich, Schnitt: Gang einer Mahlzeit
court	Platz
cover	Umschlag, Abdeckung, Deckel; zudecken
crab	Krabbe
cracked	rissig
craftsman, craftswoman	Handwerker/in
crankcase	Kurbelgehäuse
crankshaft	Kurbelwelle
crawl	klettern, kriechen
crazy	verrückt
cream-filled	mit Creme gefüllt
create	verursachen, schaffen
Creole	mittel- und südamerikanischer Mischling
crocodile clip	Klemme
croissant (fr)	Gebäck
cross	Kreuzung
cross-point screwdriver	Kreuzschraubenzieher
cross-slide	Querschlitten
cross-slot	Gegenspur
crouton (fr)	geröstete Brotstücke
crud	Schmutz
cuisine	Küche, Art zu kochen
cup	Becher, Pokal
currency	Währung
current	Stromstärke

custom-made	nach Maß-, auf Bestellung gefertigt
customer	Kunde/in
customs	Zoll
cut	schneiden; Schnitt
cut off	Abkürzung, Absperrung
cut out	Ausschnitt, Abschalter
cutting chips	(Dreh-, Bohr-)späne
cutting edges	Werkzeugschneiden
cutting tool	Schneidwerkzeug
CV (abb) curriculum vitae	Lebenslauf
cycle	Arbeitsgang
cylinder block	Zylinderblock
cylinder-pressure gauge	Flaschendruck-Manometer
cylindrical	zylindrisch
cylinder head joint	Zylinderkopfdichtung

D

dangerous	gefährlich
dare	es wagen
Darjeeling	Teesorte aus Indien
dashboard	Armaturenbrett
data	Meß- und Versuchswerte
data sheet	Datenblatt
date	Verabredung; Datum
day-return	Tagesretourfahrt
daylight-saving-time	Sommerzeit
deal (with)	sich beschäftigen, befassen (mit)
dealer	Händler, Verkäufer
debris	Trümmer, Schutt
decide	feststellen, entscheiden
decision	Entschluß, Entscheidung
declaration	Erklärung
deep-sea	Tiefsee
deficiency	Mangel
deflect	abwenden
degree	Grad
delicious	köstlich
delight	Vergnügen
deliver	liefern, bringen
delivery	(Aus-)Lieferung
demand	Anforderung, Bedarf
denomination	Nennwert
dent	Beule
deny	sperren, ablehnen
department (dept.)	Abteilung
departure	Abfahrt, Abreise
depend (on)	abhängen (von)
dependable	zuverlässig, verläßlich
deposit	deponieren, anzahlen
derailleur	(Fahrrad-)Gangschaltung
describe	beschreiben
description	Beschreibung, Darstellung
design	entwerfen, konstruieren; Entwurf, Konstruktion
design draughtsman, -woman	Konstrukteur/in
design drawing	Entwürfe
designate	kennzeichnen, bestimmen
despair	verzweifeln; Verzweiflung
despite	Trotz
destination	(Reise-)Ziel, Bestimmungsort
detect	nachweisen, entdecken
detector	Anzeigevorrichtung, Detektor
detergent	Reinigungsmittel
determine	bestimmen
develop	entwickeln
development	Entwicklung
device	Vorrichtung, Gerät
diagnostic unit	Diagnoseeinheit
dial	wählen
diameter	Durchmesser
diaphragm	halbdurchlässige Schicht, Diaphragma
die	Gewindeschneideisen
difference	Unterschied
different	verschieden
differential	unterschiedlich; Differential
difficult-to-remove	schwierig zu entfernen
dilute	verdünnen
dimension	Abmessung
dine	speisen, dinieren
direct	führen
dirt	Schmutz
dirt bike	Geländemotorrad
disabled	Behinderte/r; behindert
disadvantage	Nachteil
disappoint	enttäuschen
disassemble	zerlegen
disastrous	schrecklich, verheerend
disc	Scheibe
disc brake	Scheibenbremse
discard	ausrangieren
disconnect	abklemmen
discover	entdecken
disease	Krankheit
dish	Gericht; Geschirr
dismal	düster
dispenser	Verteiler
dispensing assistant	Apothekenassistent/in
display	zeigen, ausstellen
dispose	beseitigen, loswerden
distil	destillieren
distortion	Verformung
distributor	Verteiler
divert	ablenken, umlenken
divide	teilen
divider	Stech-, Teilzirkel
division	Abteilung
dockside	an den Docks
dog licence	Hundemarke
domestic	häuslich, einheimisch
dot	Punkt
draft	Entwurf
dramatic	dramatisch
draftsperson (US), draughtsperson, draughtsman, draughtswoman	(Konstruktions-)zeichner/in
draw	zeichnen
drawback	Nachteil
drawbar	Kupplungsstange
drawing	Zeichnung
dressing	Salatsauce
drill	bohren; Bohrmaschine
drill bit	Bohrerschneide
drilling head	Bohrkopf
drilling machine	Bohrmaschine
drive	treiben; Antrieb
driver's protection (cabin)	Sicherheitskabine
drn (abb) drawn	gezeichnet
drop in	tritt ein
droplight	herabziehbare Lampe, Zugpendellampe
drum	Trommel
dry	trocken; trocknen
dual regulators	Doppelregler
due	passend, fällig
dump	ablagern, deponieren, fallen lassen
durability	Dauerhaftigkeit, Beständigkeit
durable	beständig, dauerhaft
duration	Zeit
dust	Staub
duty	Pflicht, Abgabe

E

e.g. (abb) exempli gratia, for example	zum Beispiel
ear muffs	Ohrschützer
earth terminal	Erdanschluß

Word List

EC (abb) European Community	Europäische Union
eccentric	exzentrisch
economical	preiswert
edge	Kante, Ecke, Rand
education	Bildung, Ausbildung
efficiency	Leistungsfähigkeit, Effizienz
EGR (abb) exhaust gas recirculation	Abgasrückführung
electric drill	Elektrohandbohrmaschine
electricity	Elektrizität
electro-beam welding	Lichtbogenschweißen
electro-slag welding	Elektro-Schlackeschweißen
electrode	(Schweiß-)elektrode
element	Teil, Bandelement
eliminate	ausschalten, eliminieren
emission	Emission, Ausstoß
employer	Arbeitgeber/in
enclose	beilegen
enemy	Feind
engagement	Beschäftigung, Anstellung
engine	Motor, Maschine
engine block	Motorblock
engine tester	Motortester
engineer	Techniker/in
engineering	Technik
enjoy	sich erfreuen
ensure	sicherstellen
entangle	verfangen
entire	alles, das Ganze
entry	Anfang, Eingang
envelope	Kuvert
environment	Umwelt
environmental	die Umwelt betreffend
environmentally safe	umweltverträglich
equal	gleichwertig
equip. specs. (abb) equipment specifications	Ausrüstungsspezifikation
equipment	Einrichtung, Ausrüstung
equipped	ausgerüstet, ausgestattet
escalator	Aufzug, Lift
escape	entkommen, entfliehen
essential	notwendig, wichtig, wesentlich
established	bestehend, etabliert
estimate	beurteilen, schätzen, bewerten
evaluation	Auswertung
evaporate	verdunsten
even	regelmäßig
exact(ly)	genau, richtig
exam (abb) examination	Prüfung
examine	prüfen
example	Beispiel
excellent	ausgezeichnet
excess	Überschuß
exchange	(aus-, ein-) umtauschen
exchange rate	Wechselkurs
exclusive(ly)	ausschließlich
exhaust	auspuffen; Abgas, Auspuff
exhaust emission	Rauchgasemissionen
exhaust gas analyser	Abgasmeßgerät
exhaust valve	Auslaß- (Auspuff-)ventil
exist	vorhanden sein
expand	ausdehnen, entwickeln
expect	erwarten
expel	ausstoßen
expensive	teuer
experience	Erfahrung, Praxis
experienced	erfahren
experimental	experimentell
expert	erfahren; Fachmann, Fachfrau
expertise	Gutachten
explain	erklären
explanation	Erklärung
expression	Ausdruck
extend	verlängern, strecken
extension	Verlängerung
external	außen, extern
extract	Auszug
extraordinary	außergewöhnlich

F

face	Gesicht, Ober-, Stirnfläche
faceshield	Schutzschirm
facing	Plandrehen
fact	Tatsache
factory	Fabrik
fade in/out	ein-/ausblenden
fail	versagen
false	falsch
familiar	bekannt
family allowance	Familienbeihilfe
family run	familiengeführt
famous	berühmt
fan	Ventilator
fan wafer	Waffelröllchen
fasten	befestigen
fastener	Befestigungsmittel
fault	Fehler
FCA (abb) flux-cored wire welding	Fülldrahtschweißen
fear	fürchten
feature	Eigenschaft, Bestandteil
fee	Gebühr
feed	eingeben, füttern, zuführen
feedgear	Vorschubgetriebe
feedrod	Zugspindel, Vorschubwelle
female	weiblich
fender	Schutzblech, Kotflügel
ferry	Fähre
fibre	Faser
field tester	Testtechniker/in
fighter	Kämpfer/in
file	Ordner, Datei; Feile
file transfer	Datenübertragung
filler wire	Fülldraht
final assembly	Endfertigung
finish	fertigbearbeiten, polieren
fir	Fichte
fire extinguisher	Feuerlöscher
fireworks	Funkenregen
firm	Firma, Unternehmen; fest
fit	einpassen, einbauen, montieren
fitter	Schlosser, Monteur/in
fitting	Rohrverbindung, Kupplungsstück
fix	reparieren, befestigen
fixture	Befestigung
fizz	Sprudel
flake	Schicht, Flocke
flame-retardant	feuerfest
flammable	leicht entzündlich
flat	flach, eben; Fläche; Wohnung
flat soda pop	abgestandenes, alkoholfreies Getränk
flat-nose pliers	Flachzange
flavour	Geschmack
flexibility	Anpassungsfähigkeit
flexible	flexibel, beweglich
floor	Bodenplatte
floor model	Standmodell
floppy disc	(Computer-)Disketten
flow	fließen
flowline	Fließrichtung
fluid	Flüssigkeit
flywheel	Schwungrad
food	Speisen
foot	Fuß (0,3048 m)
foot rest	Fußraste
force	Kraft
foreign	ausländisch
foreign exchange	Geldwechsel
foreman	Vorarbeiter/in

for instance	zum Beispiel
fork	Gabel
forklift truck	Gabelstapler
form	Formular; formen
form cut	Formdrehen
formation	Aufbau, Gestaltung
formulate	formulieren
forthcoming	bevorstehend
fraction	Bruchstück
frame	Rahmen
fray	ausfransen
free	frei, umsonst
frequent(ly)	häufig
fried	gebraten
front	Vorderseite
frontload	von vorne laden
fuel	Kraftstoff
fuel cell	Kraftstoffzelle
fuel-air mixture	Kraftstoff-Luft-Gemisch
full-size	lebensgroß
full-time	Vollzeit
fume	Rauch
furnace	(Schmelz-)Ofen
further	weitere
fuse	Sicherung

G

gain	erwerben, bekommen
galvanised steel	verzinkter Stahl
gap	Lücke
gap text	Lückentext
garage	Garage, Reparaturwerkstätte
garlic	Knoblauch
garment	Kleidungsstück
gasket	Dichtung
gasoline (US)	Benzin
gateaux (franz.)	Kuchen
gauge	Manometer, Lehre; Blechdicke; (ab-, aus-)messen
gear	Getriebe; Gang
gearbox	Getriebe(gehäuse)kasten
generate	erzeugen
generator	Generator, Stromerzeuger
ginger	Ingwer
glad	glücklich
glove	Handschuh
glue	kleben; Klebstoff
GMA (abb) gas-metal arc (welding)	Schutzgasschweißen
goggles	Schutzbrille
goods	Waren, Güter
govern	steuern
grab	Griff, greifen
graduate	Absolvent/in, absolvieren
granulate	granulieren, körnen
graphic tablet	Grafiktablett
grease	Schmiermittel; schmieren
greek	griechisch
greens	Grünzeug
grid	Gitter
grind	schleifen
grinder	Schleifmaschine
grinding wheel	Schleifscheibe
grip	Handgriff
grit	grober Sand
groove	Hohlkehle, Nut, Rille
gross	brutto
ground beef	Hackfleisch
ground clearance	Bodenfreiheit
guarantee	Garantie
guide	Führer/in; Leitfaden
guide dog	Blindenhund
guide line	Leitseil, Richtlinie
gulp	Schluck; schlucken
gun	Pistole
gutter	Dach(Regen)rinne

H

H. M. (abb) Her Majesty	Ihrer Majestät
h. p. (abb) horse power	Pferdestärke
hacksaw	Bügelsäge
halibut	Heilbutt (Fisch)
halve	halbieren
ham	Schinken
hand-held filing	händischen feilen
hand-operated	händisch bedient
hand-tool	Handwerkzeug
handcart	Handkarre
handle	handhaben
handlebar	Lenkstange
handled file	Handfeile
handling	Führung, Handhabung
hands-off!	Hände weg!
handsaw	Fuchsschwanz
handset	(Telefon-)hörer
handwheel	Handrad
handy	greifbar
happen	geschehen, passieren
hard hat	Schutzhelm
harden	härten
harness	Geschirr, Gurten
hate	hassen
haul	befördern, transportieren
hazard	Gefahr
head style	(Schrauben-, Muttern-)Kopfform
headache	Kopfschmerz
headlight	Scheinwerfer
headline	Überschrift
headmaster, headmistress	Direktor/in
headstock	Spindelstock
health	Gesundheit
health reasons	gesundheitliche Gründe
heat	Hitze
heat-resistant	hitzebeständig
heatproof gloves	hitzebeständige Handschuhe
heavy	schwer
heavy commercials	Schwer-LKW
heavy duty	Hochleistungs-, Schwerlast
heel: on his last heels	Schuh-Absatz: abgerissen, heruntergekommen
height	Höhe
herbs	Kräuter
hexagonal	sechseckig
high speed steel	Schnellschnittstahl
history	Geschichte
hit	schlagen, stoßen; Treffer
hole	Loch
hollandaise (fr)	holländisch
hollow	hohl
honeycomb	wabenartig
hood (US slang for criminal)	Krimineller
hope	hoffen; Hoffnung
horizontal	waagrecht, horizontal
hose	Schlauch, Leitung
hose connector	Schlauchklemme
hose fitting	Schlauchkupplung
hose pipe joint	Schlauchverbindung
housing	Gehäuse
hub	Nabe
human	menschlich
hundredth	hundertstel
hydraulic	hydraulisch
hydride	Hydrid
hydrocarbon	Kohlenwasserstoff
hydrogen	Wasserstoff

I

identification	Identifizierung
identify	identifizieren
idle speed	Leerlauf
ignite	zünden
ignition	Zündung
illness	Krankheit

Word List

image control	Bildkontrolle
imagine	vorstellen
immediate	unmittelbar, sofort
immigration	Einwanderung
impatient	ungeduldig
important	wichtig
impressive	eindrucksvoll
improvement	Verbesserung
in-copying	Innen(kopier)drehen
inch	Zoll (25,4 mm)
incinerator	Verbrennungsanlage
include	einschließen
incomplete	unvollständig
incorporate	enthalten, miteinschließen
increase	zunehmen; vergrößern, erhöhen
incredible	unglaublich
independent	unabhängig
indicate	anzeigen
indicator	Blinklicht
indifferent	gleichgültig
inexpensive	billig
infinite(ly)	unendlich, außerordentlich
inform	unterrichten
informal	ungezwungen
infusion	Infusion (Spritze)
inhale	einatmen
inherent	eigen, inhärent
inject	einspritzen
injection	Einspritzung
injection mould	Spritzgußform
injury	Verletzung
inlet	Einlaß
input	Eingang
input area	Anlieferung
insert	einfügen
inside	innerhalb; Inneres
inside calliper	Lochtaster (-zirkel)
inside width	Innenweite (-durchmesser)
inspection cell	Prüfkammer
install	installieren
instead	anstatt
instruction	Anweisung
instructor	Lehrer/in
insurance	Versicherung
intake	Einlaß
integral	vollständig
interchangeable	austauschbar
interest	interessieren
interior	Innen(einrichtung)
internal	innen
intravenous	intravenös
intricate	kompliziert
introduce	vorstellen, bekannt machen
intruder	Einbrecher
inventory	Bestandsaufnahme, Inventar
invite	einladen
involve	einschließen
iron	Eisen
irregular	unregelmäßig
issue	Streitfrage, Ausgabe; ausstellen
item	Detail, Punkt, Artikel, Posten

J

jackhammer	Preßlufthammer
jetfoil	Tragflügelboot
jewelry	Schmuck
join	sich anschließen, verbinden; falzen
joined edges	gefalzte Kanten
joint	Verbindungsstück, Fuge
joint product	Gemeinschaftsprodukt
jolt	holpern
joy-stick	Schalthebel

K

keen	eifrig, tüchtig
key	Schlüssel
keyboard	Tastatur
kin	Verwandtschaft
kind	gut, freundlich; Art, Sorte
kinetic energy	Bewegungsenergie
kitchen sink	Abwäsche
knob	Knopf
knowledge	Wissen, Kenntnisse

L

label	Benennung, Bezeichnung, Aufschrift; bezeichnen
laboratory technician	Labortechniker/in
labour	Arbeit
lamb	Lamm
landfill	Mülldeponie
language	Sprache
large	groß
last	halten, letzte
lateral	seitlich
lathe	Drehbank
lawn-mower	Rasenmäher
lay out	zeichnen, skizzieren
layout	Plan, Skizze, Entwurf
lb (abb) pound	Pfund (453,6 Gramm)
lead	Blei; führen, leiten
leaded	verbleit
leadscrew	Leitspindel
leaflet	Prospekt
leak	Leck
least	wenigst, geringst
left-hand	linksgängig, linkshändig
left-hand thread	Linksgewinde
length	Länge
lettuce	(Kopf) Salat
lever	Hebel
licence	Zulassung, Berechtigung
license	berechtigen, zulassen
lifestyle	Lebensart
lift	heben; Aufzug
light	leicht
light commercials	Kombi und Leicht-LKW
light weight	leichtgewichtig
lighting	Beleuchtung
lightning-quick	blitzartig
line	Rohr(-leitung)
liner	Linienschiff
linger	verweilen
link	Verbindung, Bindeglied
linkage	Verbindung, Kopplung
liqueur	Likör
liquid	flüssig; Flüssigkeit
list	Verzeichnis, Liste
load	Gewicht, Last
loan	Darlehen
lobe	Zipfel, Vorsprung
lobster	Hummer
local	örtlich
locate	die Lage feststellen, liegen
location	Stelle, Lage, Platz
lock	festmachen, fixieren; Schloß
lock washer	Federring
locknut	flache Sechskantmutter
logo	(Firmen)Zeichen
long-life	langlebig
longitudinal	längs
longitudinal turning	Längsdrehen
loose	locker
lose	verlieren
loss	Verlust
lot	Geschick, Schicksal; Teilmenge
lounge	Salon
low-carbon	niedrig legiert
low-speed	niedriger Gang, langsam
Ltd (abb) Limited	Ges.m.b.H.

lubricant	Schmiermittel
lubricate	schmieren
lyrics	Liedtext

M

machine bolt	Sechskantschraube
machine fitter	Maschinenschlosser/in
machine operator	Maschinist/in
machine screw	Schrauben mit Schlitz
machine shop	Werkstätte
machine tool center	Werkzeugmaschine
machine tool operator	Werkzeugmaschinist/in
machine-screw nut	flache Sechs- oder Vierkantmutter
machinery	Maschinen, Mechanismus
machining centre	Bearbeitungszentrum
mail	Post; schicken, senden
main course	Hauptspeise
mainline	Hauptstrecke
mainly	hauptsächlich
maintain	instandhalten
maintenance	Instandhaltung
major part	Hauptbestandteil
make	Bauart
male	männlich
malt	Malz
mammal	Säugetier
manifold	Krümmer
manner	Art (und Weise)
manual	händisch
manufacture	erzeugen, herstellen
manufacturer	Hersteller
map	(Land)Karte
marine	Meeres-
marital status	Familienstand
mark	anreißen
marking tool	Anreißwerkzeug
mask	abdecken; Abdeckung
mass	Masse, Menge
match	entsprechen, anpassen
materials	Stoffe, Materialien
maths	Mathematik
matter	Angelegenheit, von Bedeutung sein
maximum speed	Spitzengeschwindigkeit
meal	Essen, Speise
means	mittels; Mittel
meantime: in the meantime	Zwischenzeit: unterdessen
measles	Masern
measure	Maßregel: ausmessen
measurement	Abmessung
mechanic	Mechaniker/in
mechanical	mechanisch
mechanical engineer	Maschinenbauingenieur/in
mechanical engineering	Maschinenbau
mechanical fitter	Maschinenschlosser/in
medicare	medizinische Betreuung
meet	treffen
melt	schmelzen
member	Mitglied
membership	Mitgliedschaft
memory	Speicher
mention	erwähnen
mercury	Quecksilber
metal gauge	Blechlehre
metal stock	Metallvorrat
metallurgy	Metallurgie
metalwork	Metallarbeit
meticulous	übergenau
metric	metrisch
micrometer	Mikrometerschraube
MIG (abb) metal inert gas	Metall Inert Gas
MIG welding	Schutzgasschweißen
mill	drehen, fräsen
milling machinist	Dreher/in
milling tool	Dreh/Fräswerkzeug
minced meat (GB)	Faschiertes
miner	Bergmann
mint	Minze
mist	Nebel
mistake	Irrtum, Fehler
mixture	Mischung
mixture setting	Gemischeinstellung
modernize	modernisieren
modest	bescheiden, maßvoll
modify	abändern, modifizieren
modulate	regulieren
module	Verhältnismaß, Modul
mold (US)	(Guß)Form; formen, gießen
molten	geschmolzen
monitor	Bildschirm, Monitor
monoxide	Monoxyd
motor frame	Motorrahmen
motor mechanic	Automechaniker/in
mould (GB)	(Guß)Form; formen, gießen
mount	montieren, Träger
mounted	montiert, zusammengebaut
move	(die Firma) wechseln, bewegen
movement	Bewegung
mud	Schlamm
muffler	Auspufftopf
mushroom	eßbarer Pilz
musician	Musiker/in
mussels	(Mies)Muscheln

N

nail punch	Durchschlag
nationality	Nationalität
nature	Art
neat	sauber, ordentlich
necessary	notwendig
need	brauchen
negotiable	überwindbar, passierbar
neighbour	Nachbar/in
nervous	nervös
nitrogen	Stickstoff
noise	Lärm
noisy	geräuschvoll
non-slip	rutschfest
nook	Winkel, Ecke
nose, nosing	Nase, hakenförmiger Ansatz
notch	Nut
notebook	Notizbuch
notice	bemerken; Ankündigung
noun	Hauptwort
nozzle	Düse
number	Zahl
numerical	numerisch
numerous	zahlreich
nursery school	Kindergarten, Vorschule
nursing home	Pflegeheim
nut	(Schrauben-)Mutter
nut-driver	Schraubenzieher

O

OAP (abb) old age pensioner	über 60. bzw. 65 Jahre alte Menschen
object	Gegenstand, Zweck
observe	beachten
obstacle	Hindernis
obtain	erreichen, erhalten
occupation	Beruf
occupational	beruflich
occur	vorkommen
odd	ausgefallen, nicht dazupassend
offer	Angebot: anbieten
office	Büro
oil filter	Ölfilter
onion	Zwiebel
operate	arbeiten
operating panel	Bedieneinheit
operating speed	Betriebsgeschwindigkeit

Word List

operation	Betrieb
operator	Arbeiter/in, Telefonist/in
opportunity	Gelegenheit
opposite	gegenüberliegend
optimize	optimieren
option	Möglichkeit
order	Reihenfolge; bestellen
out-copying	Außen-(kopieren)drehen
outer	äußere(r, s)
outfit	Ausrüstung
outlet valve	Auslaßventil
outline	anreißen
output	Ergebnis; Produktion
outside	außerhalb
outside calliper	Außentaster
outside micrometer	(Außen-)Mikrometerschraube
over-squeeze	Überdruck
overall	gesamt
overhead	über Kopf, obere(r, s)
overlap	überlappen
own	besitzen
oxidation, oxide	Oxydation, Oxyd, Sauerstoffverbindung
oxidize	oxidieren
oxyacetylene	Sauerstoff-Azetylen ...
oxyacetylene gas welding	Autogenschweißen
oxygen	Sauerstoff
oyster	Auster
P	
pack(age)	Paket
packaging	Verpackung
page	Seite
pain (fr)	Brot
paint	lackieren; Lackierung, Farbe
paint sprayer	(Auto-)Lackierer/in
pallet	Palette
pan	Wanne
panel	Tafel; Blech
paragraph	Absatz
part way (half way)	halbwegs
part-time	Halbtags-, Aushilfsarbeit
particular	besonders
part off	(ab)trennen
parting tool	Trennwerkzeug
pass	bewegen
passage	Absatz
passive voice	Leideform
pastrami	geräuchertes Rindfleisch
pasture	Weide
patent	offenbar, patentiert
path	Weg
pay	Lohn
peak	Gipfel
peasant	Bauer/in
pecan	Hickorynuß
peeling action	Schälvorgang (Zerspanen)
peen	Pinne, Hammerbahn
perform	ausführen
performance	Vorstellung, Leistung
permanent seam	Dauernaht
permissible	zulässig
permit	Zulassung, Lizenz
perpendicular	senkrecht
personal data	Angaben zur Person
petrol (GB)	Benzin
phone call	Telefonanruf
photovoltaic	photoelektrisch
phrase	Redensart
physical	körperlich
physics	Physik
pick	auswählen, wegnehmen
pick up	Anstieg, Erholung; aufheben
pie	Pastete
piece	Stück
pillar	Säule, Ständer
pillar drill (press)	Säulenbohrmaschine
pincers	Beiß-,Zwickzange
pineapple	Ananas
ping	Klingeln (des Motors)
pipe	Rohr
pipe fitter	Gas- und Wasserinstallateur/in
piston	Kolben
pitch	Ganghöhe, Steigung
pitch gauge	Blechlehre
plan	(Maß)Zeichnung, Riß
planetary grinding	Planetenschleifen
plant	(Betriebs)Anlage, Werk, Fabrik
plant layout	Fabrikslageplan
plasticize	ins Plastische verformen
plate	Deckel: Platte
pleasant	angenehm
pleased	erfreut
pliers	Zange
plot	eine Linie zeichnen
plow (US), plough (GB)	Pflug
plug	Pfropfen, Bolzen
pneumatic	pneumatisch
point	nach einer Richtung weisen
political education	Politische Bildung
pollution	Umweltverschmutzung
pork	Schweinefleisch
portable	tragbar
porter	Portier
position	Position, Stelle
possibility	Möglichkeit
possible	möglich
post	Pfosten, Ständer
postal order	Postanweisung
postal rate	Postgebühr
pot	Topf
potential	Möglichkeit
pour	Guß; gießen
powder	Pulver
power	Energie, Kraft
power cord	Stromkabel
power drill	elektrische (Hand)Bohrmaschine
powerful	stark, kräftig
power plant	Kraftwerk
practical	wirklich
practice	üben, trainieren
practise	Praxis
prawn	Garnele
precision	Präzision, Genauigkeit
predominant	vorherrschend
prefer	bevorzugen
prejudice	Vorurteil; beeinträchtigen, schädigen
premium	Bonus, Prämie, Zuschlag
prepare	vorbereiten, ausarbeiten
pressure	Druck
prev. bal (abb) previous balance	Übertrag
prevent	verhindern
previous (experience)	vorher; Vor-(erfahrung)
price tag	Preisschild
pride	Stolz
primarily	hauptsächlich, in erster Linie
prime	erstklassig
principal (US)	Direktor/in
print	technische Zeichnung
print reading	technische Zeichnungen lesen
printed matter	Drucksache
printer	Drucker
probably	möglicherweise
probe	Sonde; sondieren
process	bearbeiten, ausstellen
produce	produzieren, erzeugen
product line	Fließband, Fertigungsstraße
product presentation	Produkt(Werbe)vorführung
production technique	Fertigungstechnik
production unit	Fertigungseinheit
productivity	Produktivität

profession	Beruf	**receive**	erhalten, empfangen
professional	Profi	**recent**	neulich
professional magazine	Fachzeitschrift	**reception**	Empfang
profile	Querschnitt	**receptionist**	Empfangsdame/herr
programmer	Programmierer/in	**recess**	Vertiefung
progressive	fortschrittlich	**recognize**	erkennen
project	Projekt, Plan; planen	**recommend**	empfehlen
promotion	Beförderung	**record**	aufnehmen, niederschreiben
prop stand	Ständer	**recruitment**	(Personal)Aufnahme
propel	treiben	**recur**	wiederkehren
proper	richtig, ordentlich, vorschriftsmäßig	**recycle**	wiederverwenden
property	Besitz, Eigenschaft	**reduce**	reduzieren
proportion	Verhältnis	**redundant**	überflüssig
protect	schützen, verkleiden	**reel**	Rolle, Spule
protection	Schutz	**reference**	Hinweis, Bezug nehmen auf
protection technique	Schutztechnik	**reflect**	reflektieren
protective	schützend	**refrigerator**	Kühlschrank
prototype	Prototyp, Modell	**refurbishment**	Renovierung
prove	beweisen, nachweisen	**regards**	mit herzlichen Grüßen
provide	sorgen	**register**	sich anmelden, eintragen
public	öffentlich	**registration**	Anmeldung, Eintragung
publish	veröffentlichen	**regulate**	regulieren
publisher	Verleger/in	**regulation**	Anordnung
punch	Kraft, Energie, Schlag; schlagen	**regulator**	Regler, Reguliervorrichtung
punctual	pünktlich	**relate**	beziehen
purchase	kaufen	**relax**	entspannen
pure	rein	**release**	abgeben
purge	reinigen	**relevant**	anwendbar
purpose	Zweck	**reliable**	verläßlich
push	drücken	**relocation**	Umsiedlung
put on	anziehen	**remarkable**	bemerkenswert
		remove	entfernen
		renovate	renovieren
Q		**renowned**	berühmt
quality	Qualität	**repair**	Reparatur; reparieren
quality controller	Qualitätsprüfer/in	**repair shop**	Reparaturwerkstätte
quantity	Quantität, Menge	**repeat**	wiederholen
quarter (turn)	Viertel(drehung)	**replace**	ersetzen
question	Frage	**reply**	Antwort
quick-change	Schnellwechsel	**report**	Bericht, Zeugnis
quiet	still, ruhig	**report grade**	Zeugnisnote
quite	ziemlich, wirklich, durchaus	**representative**	Vertreter; repräsentativ
quote	zitieren	**reputation**	Ansehen, Ruf
		require	(er)fordern, benötigen
		research	Forschung
R		**resistance**	Widerstand
r. p. m. (abb) revolutions per minute	Umdrehungen pro Minute	**resistant**	widerstandsfähig
		resistor	elektrischer Widerstand
rack	Zahnstange	**respective**	jeweilig
radial	senkrecht zur Längsachse	**respiration**	Atmung
radial arm drill	Radialbohrmaschine	**respirator**	Atemgerät, Gasmaske
radiator	Kühler	**responsibility**	Verantwortung
radius (Mz. radii)	Halbmesser	**restrict**	einschränken, beschränken
rag	Fetzen	**result**	Ergebnis
rail	Geländer, Schiene	**retard**	hemmen
raise countersunk	Linsensenkkopf	**retirement**	Ruhestand
range	Umfang, Bereich; reichen	**retrieve**	(wieder) auffinden
rapid	rasch	**return**	zurückgeben
rare	selten	**reuse**	wiederverwenden
raster	Raster	**rev (abb) revolution**	drehen; Umdrehung
ratchet	Ratsche, Sperrklinke	**reverse**	zurück, retour, umpolen
rate	Rate, Gebühr, Kurs	**reversible**	umkehrbar, umsteuerbar
rather	ziemlich	**revolve**	drehen, rotieren
rating (power rating)	(elektrische) Leistung	**reward**	Belohnung; belohnen
ratio	Verhältnis	**rib**	Rippe
rationalization	Rationalisierung	**ride**	gleiten
raw (material)	Roh-(stoff)	**rife**	häufig, verbreitet
re-usable	wieder verwendbar	**right-angled**	rechtwinkelig
reach	erreichen	**right-handed**	rechtshändig
reaction	Reaktion	**right-hand thread**	Rechtsgewinde
real	wirklich, real	**rigid**	starr, steif
realise	verwirklichen	**rim**	Felge, Radkranz
rear	hintere Seite	**rinse**	spülen
reason	Grund	**rise**	(an)steigen
reasonable	annehmbar	**risk**	Risiko; riskieren
rebuild	umbauen, wiederherstellen	**roast**	rösten
receipt (of)	Eingang (von)	**rod**	Stange

Word List

roll feed	Vorschub
roller	Rolle
roller-bearing	Rollenlager
rolls	(Waffel)Röllchen
rotary movement	Drehmoment
rotate	rotieren, drehen
rotation	Drehung
rough(ly)	ungefähr
round	rund
row	Reihe
royalty	Königshaus
rpm (abb) revolutions per minute	Umdrehungen pro Minute
rub	reiben
rubber	Gummi
rubber backed disc	Gummischeibe
rubbish	Abfall
rude	unhöflich
rule	Maßstab
rumour	Gerücht
run	laufen
rush	schießen, stürzen von Wasser
rye	Roggen

S

saddle	Schlitten
safe	sicher
safety	Sicherheit
salary	Gehalt
sale	Verkauf, Ausverkauf
sales assistant	Verkäufer/in
sample	Muster, Probe
satisfaction	Befriedigung
satisfactory	zufriedenstellend, befriedigend
save	sichern, sparen
savings deposit	Spareinlage
saw	Säge; sägen
saw rasp	Raspel
scale	Maßstab
scanner	Scanner
scene	Schauplatz, Ansicht
schedule	Tabelle
schematic	schematisch
science	Wissenschaft
scone	Brötchen aus Rührteig
score	Ergebnis, Punktezahl
Scotch	schottisch
scratch	Kratzer; ritzen
scream	quietschen
screen	Bildschirm
screen-based	bildschirmorientiert
screw	Schraube
screw conveyor	Förderschnecke
screwdriver	Schraubenzieher (-dreher)
scribe	anreißen
scriber	Reißnadel
script	Schrift(art); Manuskript
seal	versiegeln, abdichten; Abdichtung
sealant	Dichtmittel
seam	Falz, Naht, Fuge; falzen
seat	Sitz
second-hand	gebraucht
section	Teil, Abschnitt
secure	sichern
security	Sicherheit, Schutz
seed	Kern
seek	anstreben, begehren
seem	scheinen
select	auswählen
self-assessment	Selbstbewertung
self-drilling	selbstbohrend
self-grip	selbsthaltend
self-oiling	selbstschmierend
self-tapping	selbstgewindeschneidend
self-reliant	selbständig
sell	verkaufen
semi	halb
semi-automatic	halbautomatisch
semiskilled	halbausgebildet
send	senden, schicken
sense	Sinn, Verstand
sensible	vernünftig
sensor	Fühler
sentence	Satz
separate	trennen, getrennt
sequence	Reihenfolge
serious	ernsthaft
service	Dienst
service technician	Servicetechniker/in
service weight	Leergewicht
service tool	Reparaturwerkzeug
set	Sortiment, Satz von
set bolt	Kopfbolzen
set-up	Anlage
setting	Einstellung
several	mehrere, einige
sewage	Abwasser
shaft	Welle
shake	schütteln
shank	Schaft, Stiel
shape	Form; formen
shaping	kurzhobeln (Metall)
share	teilen
shaver	Rasierer
sheet	dünne Platte, Blatt
sheet metal worker	Spengler/in
sheet-metal	Blech
shell	Schale; Außenhaut, Verkleidung
shield	schützen
shift	(um)schalten
shift gear	Schaltgetriebe
shock absorber	Stoßdämpfer
shop	Werkstätte, Abteilung einer Fabrik
shoulder	An-, Absatz, Schulter, Stufe
show	zeigen; Vorstellung
shower	Dusche
showroom	Ausstellungsraum
shutdown	Stillegung
shy	schüchtern
sickleave	Krankenurlaub
side order	Beilage
sign	Zeichen
signal	Signal; signalisieren
signature	Unterschrift
silencer	Schalldämpfer
silicone rubber	Silicongummi
similar	ähnlich
simple	einfach
simultaneous	gleichzeitig
single	einzeln, einfach, allein
single room	Einzelzimmer
sink	Abwasch
sirloin	Lendenstück
site	Baustelle
situation	(Arbeits)Stelle
size	Größe
skill	Fertigkeit
skilled worker	Facharbeiter/in
slaughter	Schlachtung
sleeve	Hülse, Buchse, Muffe
slice	Scheibe
slide	gleiten
sliding bevel	Schmiege, Winkelpasser
slight	leicht
slot	Nut
sluice	Schleuse
small	schmal, klein
smell	Geruch, Gestank
smog	Smog
smoke	Rauch
smooth	glätten
smudge	Schmutzfleck
snow guard	Schneefänger

society	Gesellschaft
soft faced hammer	Kunststoffhammer
software	Computerprogramme
solar power	Sonnenenergie
solder	löten
solder wire	Lötzinn, Lötdraht
soldering iron	Löteisen, Lötkolben
solenoid	Solenoid
solid	massiv
solid steel	Quadratstahl
solidify	erstarren
solution	Lösung
solve	lösen
solvent	Lösungsmittel
sort	Art, Sorte
sound	gesund, korrekt; Schall
sour cream	Sauerrahm
source	Quelle
social security	Sozialversicherung
space	Platz, Weltraum
space-age	Raumzeitalter
spanner	Schraubenschlüssel
spare	Reserve
spark	Funke, Zündfunke
spark plug	Zündkerze
speciality fields	Spezialfächer
specific	bestimmt
specification	Spezifikation
specify	spezifizieren
speed	Geschwindigkeit
spell	buchstabieren
spend	auf-, verwenden
spherical	kugelförmig
spillway	Überlaufrinne
spin	drehen, schleudern
spin-casting	Schleuderguß
spinach	Spinat
spindle	Spindel
spindle keys	Spindelschlüssel
spire	Kirchturm
spirit level	Wasserwaage
splash	spritzen
spoke	Speiche
spokesman	Sprecher/in
spontaneously	spontan
spool	Spule
spot welding	Punktschweißen
spray	spritzen
spray paint	Spritzlackierung
spraypainter	Lackierer/in
spread	ausbreiten
spring	Feder
sprocket	Zahnkranz
square	Platz; quadratisch, rechtwinkelig machen, vierkantig formen
square collar	quadratische Manschette, Quadratstahl
square nut	Vierkantmutter
squareness	quadratische od. rechteckige (Form)
squeeze	Druck
stabilization	Festigung
staff	Belegschaft, Personal
stage set	Bühnenbild
stain	Fleck
stainless steel	rostfreier Stahl
stake	Faust, Stöckel
stamp	Briefmarke
standard	Norm
state	angeben
state-of-the-art	auf dem letzten Stand (der Technik)
statement	Aussage, Erklärung, Behauptung
stationary	stationär
statistical	statistisch
steady	gleichmäßig
steel	Stahl
steel centre punch	Körner
steel rule	Stahlmaßband, Stahllineal
steel square	Stahlwinkel
steel-tipped divider	Stechzirkel mit gehärteten Spitzen
steep	steil
steer	steuern
steering	Lenkung
step	Schritt; abschreiten
stick	Stock; stecken
still	ruhig; noch immer
stock	Vorrat
stoichiometric	stöchiometrisch
storage	Lager, Depot
storage box	Kleinteilebox
store	speichern, lagern
strawberry	Erdbeere
stress	betonen, unterstreichen
strip	(Blech-)Streifen
stroke	Takt
strong	stark, tüchtig
structure	Konstruktion, Gefüge, Bauart
stud	Stift
stud bolt	Schraubenbolzen
study	studieren
sturgeon	Stör (Fisch)
stylish	elegant, stilvoll
sub frame	Hilfsrahmen
sub-assembly	Zwischenfertigung
subject to	vorbehaltlich
subject	Unterrichtsgegenstand
subsequent	nachträglich, später
subsidiary	Tochtergesellschaft
successful	erfolgreich
suck	saugen
suit	Anzug
suitable	geeignet
sump	(Öl)sumpf
sump drain plug	Ölablaßschraube
supply	Versorgung; versorgen
support	abstützen, halten; Unterlage
suppose	annehmen
surface	Oberfläche
surface grinder	Flächenschleifmaschine
surface letter	auf Land- oder Wasserweg zugestellter Brief
survey	Übersicht, Umfrage
suspend	aufhängen
suspension	Aufhängung, Federung
sweets	Süßigkeiten
swingarm	Federgabel
switch	Schalter
synthesize	synthetisch, künstlich bilden

T

table d'hote (fr)	Wirtstafel
tailrace	Endlauf
tailstock	Reitstock
take	Probeaufnahme; nehmen
take-off	Start
tall	groß
tap	Wasserhahn; Gewindebohrer, erschließen
taper	zuspitzen
task	Aufgabe, Ziel
taste	Geschmack
tax	Steuer
teach	lehren, unterrichten
technical	technisch
technician	Techniker/in
technique	Verfahren
technology	Technik
telecommunication	Fernmeldetechnik
telephone directory	Telefonbuch
teleprinter	Fernschreiber
template	Schablone
temporary	zeitweise
ten-speed bike	Zehngangrad

Word List

term	Zeit, Frist
terminal	Pol; Anschlußklemme
test	Prüfung; prüfen
tester	Prüfer/in; Testvorrichtung
thermal	thermisch
thermoplastic	Thermoplast
thick	dick
thickness	Dicke
thimble	Zwinge
thin	dünn
thin nut	dünne Vierkantmutter
though	jedoch, obwohl
thorough	gründlich
thousandth	Tausendstel
thread	Gewinde
threat	Gefahr
three-jaw chuck	Dreibackenfutter
three-phase	Dreiphasen
three-way	Dreiweg...
thrill	Aufregung
throttle	Drossel(ventil)
throw	werfen
thumbscrew	Flügelschraube
tick	abhaken
ticket window	Fahrkartenschalter
tie	binden
tight	fest(sitzend)
tight-fitting lid	dicht schließender Deckel
tile	(Dach)Ziegel
time-off	Freizeit, Freistellung
timetable	Fahrplan, Stundenplan
timing	zeitliche Steuerung
timing light	Zündfolgebestimmung
tin-lead	Lötzinn
tin-snips	Blechschere
tiny	klein
tip	Düse
titanium	Titan
together	zusammen
tolerance	Nachsicht; Abmaß, Abweichung
toll-free	zollfrei, gebührenfrei
tomorrow	morgen
toner	Toner
tool	Werkzeug
tool kit	Werkzeugausstattung
tool post	Schneidstahlhalter
tool-and-die maker	Werkzeugmacher/in
tool-rest	Oberschlitten
toolmaker	Werkzeugmacher/in
toothpaste	Zahnpaste
top hose	oberer Schlauch
topic	Thema
torch	Brenner
torch fitting	Brennerkupplung
torque	Drehmoment
touch	Berührung, Druck
toward	nach ... zu, gegen ... hin
toxic	giftig, toxisch
toy	Spielzeug
track	(Fahr)spur; Spur(weite)
track gauge	Spurweitenlehre
track width	Spurweite
tractive effort	Zugkraft
tractor	Traktor
trade	Handel
trademark	Warenzeichen
tradesman	Handwerker
trading report	Geschäftsbericht
traffic	Verkehr
train	ausbilden
trainee	Anlernling
trammel points	Stangenzirkel
transducer	Umwandler
transfer	übertragen
translate	übersetzen
transmission	Getriebe, Übersetzung, Leitung
transport	befördern; Transport
trash	Abfall
traveller	Reisende/r
travellers cheque	Reisescheck
tread circle	Kurvenradius
trepidation	Angst
trial balance	Probebilanz
tricky	kompliziert, schwierig
trigger switch	Druckschalter
trim	zurichten, herrichten
trimmer	Zurichter/in
triple	dreifach
trolley	Förderkarren, Oberleitungsbus
trouble	Schwierigkeiten, Kummer, Mühe
trouble-free	problemlos
truck	LKW
trumpet	Trompete
try square	Anschlagwinkel
tube	(Leuchtstoff)Röhre
turbocharger	Turbolader
turkey	Truthahn
turn	drehen, antreiben; Umdrehung
turn off	ausschalten
turned-up	aufgebogen
turner/machinist	Dreher/in
turning	Drehen
turret	Revolver (Drehmaschine)
two-axis	Doppelachse
two-speed	Zweigang
two-wheeled	zweirädrig
type	Art
typist	Maschinschreibkraft
tyre (GB), tire (US)	Reifen

U

UK (abb) United Kingdom	Vereinigtes Königreich
ultrasonic	Ultraschall
ultrasonic welding	Ultraschallschweißen
unattended	unbeaufsichtigt
unbeatable	unschlagbar
undercarriage	Fahrgestell, Rahmen
underline	unterstreichen
undo	entfernen
unfamiliar	unbekannt, fremd
union	Gewerkschaft
unique	einmalig
unit	Einheit
unitary skeleton	Schalenbauweise
universal cutter	Universalschneider
unleaded	unverbleit
unlimited	unbeschränkt
unmanned	unbemannt, leer
unproportionate	unproportional
unscrew	abschrauben
unthreaded	ohne Gewinde
up-and-down	auf und ab
up-to-date	modern
up-to-the-minute	auf die Minute genau
upholstery	Polsterung
upright	senkrecht
upright drill press	Ständerbohrmaschine
upstanding	aufrecht stehend
upturn	umdrehen
use	benützen
useful	nützlich
user	Anwender
usual	üblich
usually	normalerweise
utilize	nutzen

V

V-belt	Keilriemen
vacancy	freie Stelle
vacant	frei
vacation (US)	Ferien
value	Wert

valve	Ventil
van	Kombi, Kleinbus, Wagen
vapor	Dampf, Gas
vaporise	vergasen
vapour	Gas
variable	veränderlich, unbeständig, wechselnd
variety	Vielfalt, Auswahl
various	verschieden
vary	abändern, abweichen
vast	groß
vegetable	Gemüse
vegetarian	Vegetarier/in
vehicle	Fahrzeug
veil	Schleier
vent	Loch, Öffnung für Lüftung
vent hole	Luftloch
ventilation	Lüftung
verb	Zeitwort
vernier calliper	Schiebelehre
vertical	senkrecht
via	über
vice	Schraubstock
vicinity	Nähe, Nachbarschaft
virtual	tatsächlich
virtually	im Grunde genommen, im wesentlichen
viscosity	Viskosität, Zähflüssigkeit
visible	sichtbar
visit	besuchen
visitor	Besucher/in
visual	visuell
vocational	beruflich
vocational school	Berufsschule
voice	Stimme
voltage	elektrische Spannung

W

wafer	Waffel
wafer stick	Hohlwaffel
wage	Lohn
waiter/waitress	Kellner/in
wall	Mauer, Fassade
wallboard	Baupappe
warn	warnen
washer	Beilagscheibe
waste	Müll
waste gas	Auspuffgas, Abgas
watch	beobachten; Uhr
water pump	Wasserpumpe
way	Weg, Richtung
wear	(Kleider) tragen; Abnützung
weather	Wetter
weathering	Abdeckung
weigh	wiegen
weight	Gewicht
weld	schweißen
welder	Schweißer/in
well presented	gut aussehend
Welsh	Waliser/in
wet or dry paper	Naß- oder Trockenschleifpapier
wetness	Nässe
whale	Wal
wheel	Rad
wheel arch	Radkasten
wheel arrangement	Radanordnung
wheel base	Radstand
wheel bearing	Radlager
wheel rim	Felge
wheelbase	Rad(ab)stand
wheelchair	Rollstuhl
wheelset	Radsatz
while	während
whip	schlagen
wholemeal	Vollkorn
wide	breit, groß
width	Breite
wing	(Kot)Flügel
wing nut	Flügelmutter
wipe out	ausrotten
wire	Draht
wire brush	Drahtbürste
wiring	Verkabelung, Verdrahtung
withhold	zurückhalten
without	ohne
wood	Wald, Holz
work	Arbeit; arbeiten
work surface	Arbeitsfläche
workbench	Werkbank
workforce	Belegschaft
working-pressure gauge	Arbeitsdruck-Manometer
workpiece	Werkstück
workplace	Arbeitsplatz
workshop	Werkstatt
workwear	Arbeitskleidung
worn	abgenutzt
worry	beunruhigen
worth	wert
wrap	(ein)wickeln
wrapper	Verpackung, Papier
wrench	Schraubenschlüssel
wrong	falsch
wrought-iron	Schmiedeeisen

Y

year	Jahr
Yellow Pages	Branchenverzeichnis (Telefonbuch)
yield	Ertrag
youngster	junger Mensch

Z

zero	Null
zip (code)	Postleitzahl